tredition

tredition was established in 2006 by Sandra Latusseck and Soenke Schulz. Based in Hamburg, Germany, tredition offers publishing solutions to authors and publishing houses, combined with worldwide distribution of printed and digital book content. tredition is uniquely positioned to enable authors and publishing houses to create books on their own terms and without conventional manufacturing risks.

For more information please visit: www.tredition.com

TREDITION CLASSICS

This book is part of the TREDITION CLASSICS series. The creators of this series are united by passion for literature and driven by the intention of making all public domain books available in printed format again - worldwide. Most TREDITION CLASSICS titles have been out of print and off the bookstore shelves for decades. At tredition we believe that a great book never goes out of style and that its value is eternal. Several mostly non-profit literature projects provide content to tredition. To support their good work, tredition donates a portion of the proceeds from each sold copy. As a reader of a TREDITION CLASSICS book, you support our mission to save many of the amazing works of world literature from oblivion. See all available books at www.tredition.com.

 Project Gutenberg

The content for this book has been graciously provided by Project Gutenberg. Project Gutenberg is a non-profit organization founded by Michael Hart in 1971 at the University of Illinois. The mission of Project Gutenberg is simple: To encourage the creation and distribution of eBooks. Project Gutenberg is the first and largest collection of public domain eBooks.

Art in Needlework A Book about Embroidery

Lewis Foreman Day

Imprint

This book is part of TREDITION CLASSICS

Author: Lewis Foreman Day
Cover design: Buchgut, Berlin – Germany

Publisher: tredition GmbH, Hamburg - Germany
ISBN: 978-3-8472-2300-9

www.tredition.com
www.tredition.de

Copyright:
The content of this book is sourced from the public domain.

The intention of the TREDITION CLASSICS series is to make world literature in the public domain available in printed format. Literary enthusiasts and organizations, such as Project Gutenberg, worldwide have scanned and digitally edited the original texts. tredition has subsequently formatted and redesigned the content into a modern reading layout. Therefore, we cannot guarantee the exact reproduction of the original format of a particular historic edition. Please also note that no modifications have been made to the spelling, therefore it may differ from the orthography used today.

ART IN NEEDLE-WORK

TEXT-BOOKS OF ORNAMENTAL DESIGN

ART IN NEEDLEWORK

A BOOK ABOUT EMBROIDERY
BY
LEWIS F. DAY
AUTHOR OF 'WINDOWS,' 'ALPHABETS,'
'NATURE IN ORNAMENT' AND OTHER
TEXT-BOOKS OF ORNAMENTAL DESIGN
& MARY BUCKLE
LONDON:
B. T. BATSFORD 94 HIGH HOLBORN
1900
BRADBURY, AGNEW, & CO. LD., PRINTERS,
LONDON AND TONBRIDGE.

PREFACE. [v]

Embroidery may be looked at from more points of view than it would be possible in a book like this to take up seriously. Merely to hover round the subject and glance casually at it would serve no useful purpose. It may be as well, therefore, to define our standpoint: we look at the art from its practical side, not, of course, neglecting the artistic, for the practical use of embroidery is to be beautiful.

The custom has been, since woman learnt to kill time with the needle, to think of embroidery too much as an idle accomplishment. It is more than that. At the very least it is a handicraft: at the best it is an art. This contention may be to take it rather seriously; but if one esteemed it less it would hardly be worth writing about, and the book, when written, would not be worth the attention of students of embroidery, needleworkers, and designers of needlework to whom it is addressed. It sets forth to show what decorative stitching is, how it is done, and what it can do. It is illustrated by samplers of stitches; [vi] by diagrams, to explain the way stitches are done; and by examples of old and modern work, to show the artistic application of the stitches.

A feature in the book is the series of samplers designed to show not only what are the available stitches, but the groups into which they naturally gather themselves, as well as the use to which they may be put: and the back of the sampler is given too: the reader has only to turn the page to see the other side of the stitching—which to a needlewoman is often the more helpful. Lest that should not be enough, the stitches are described in the text, and a marginal note shows at a glance where the description is given. This should be read needle and thread in hand—or skipped. Samplers and other examples of needlework are uniformly on a scale large enough to show the stitch quite plainly. The examples of old work illustrate always, in the first place, some point of workmanship; still they are chosen with some view to their artistic interest.

In other respects Art is not overlooked; but it is Art in harness. Design is discussed with reference to stitch and stuff, and stitch and stuff with reference to their use in ornament. It has been endeavoured also to show the effect needlework has had upon pattern, and the ways in which design is affected by the circumstance that it is to be embroidered.

The joint authorship of the work needs, perhaps, [vii] a word of explanation. This is not just a man's book on a woman's subject. The scheme of it is mine, and I have written it, but with the co-operation throughout of Miss Mary Buckle. Our classification of the stitches is the result of many a conference between us. The description of the way the stitches are worked, and so forth, is my rendering of her description, supplemented by practical demonstration with the needle. She has primed me with technical information, and been always at hand to keep me from technical error. With reference to design and art I speak for myself.

My thanks are due to the authorities at South Kensington for allowing us to handle the treasures of the national collection, and to photograph them for illustration; to Mrs. Walter Crane, Miss Mabel Keighley, and Miss C. P. Shrewsbury, for permission to reproduce their handiwork; to Miss Argles, Mrs. Buxton Morrish, Colonel Green, R.E., and Messrs. Morris and Co., for the loan of work belonging to them; and to Miss Chart for working the cross-stitch sampler.

I must also acknowledge the part my daughter has had in the production of this book: without her constant help it could never have been written.

LEWIS F. DAY.

January 1st, 1900.

ART IN NEEDLEWORK. [1]

EMBROIDERY AND STITCHING.

Embroidery begins with the needle, and the needle (thorn, fishbone, or whatever it may have been) came into use so soon as ever savages had the wit to sew skins and things together to keep themselves warm—modesty, we may take it, was an afterthought—and if the stitches made any sort of pattern, as coarse stitching naturally would, that was *embroidery*.

The term is often vaguely used to denote all kinds of ornamental needlework, and some with which the needle has nothing to do. That is misleading; though it is true that embroidery does touch, on the one side, *tapestry*, which may be described as a kind of embroidery with the shuttle, and, on the other, *lace*, which is needlework pure and simple, construction "in the air" as the Italian name has it.

The term is used in common parlance to express any kind of superficial or superfluous ornamentation. A poet is said to embroider the truth. [2] But such metaphorical use of the word hints at the real nature of the work—embellishment, enrichment, *added*. If added, there must first of all be something it is added *to*—the material, that is to say, on which the needlework is done. In weaving (even tapestry weaving) the pattern is got by the inter-threading of warp and weft. In lace, too, it is got out of the threads which make the stuff. In embroidery it is got by threads worked *on* a fabric first of all woven on the loom, or, it might be, netted.

There is inevitably a certain amount of overlapping of the crafts. For instance, take a form of embroidery common in all countries, Eastern, Hungarian, or nearer home, in which certain of the weft threads of the linen are *drawn out*, and the needlework is executed upon the warp threads thus revealed. This is, strictly speaking, a sort of tapestry with the needle, just as, it was explained, tapestry itself may be described as a sort of embroidery with the shuttle. That will be clearly seen by reference to Illustration 1, which shows a fragment of ancient tapestry found in a Coptic tomb in Upper Egypt. In the lower portion of it the pattern appears light on dark.

As a matter of fact, it was wrought in white and red upon a linen warp; but, as it happened, only the white threads were of linen, like the warp, the red were woollen, and in the course of fifteen hundred years or so much of this red wool has perished, leaving the [4] white pattern intact on the warp, the threads of which are laid bare in the upper part of the illustration.

larger image

1. TAPESTRY, SHOWING WARP.

It is on just such upright lines of warp that all tapestry, properly so called, is worked—whether with the shuttle or with the needle makes no matter—and there is good reason, therefore, for the name of "tapestry stitch" to describe needlework upon the warp threads only of a material (usually linen) from which some of the weft threads have been *withdrawn*.

The only difference between true tapestry and drawn work, an example of which is here given, is, that the one is done on a warp that has not before been woven upon, and the other on a warp from which the weft threads have been *drawn*. The distinction, therefore, between tapestry and embroidery is, that, worked on a warp, as in Illustration 1, it is tapestry; worked on a mesh, as in Illustration 3, it is embroidery.

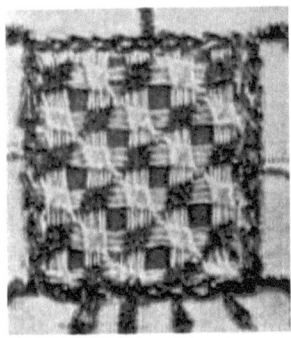

larger image

2. DRAWN WORK.

With regard, again, to lace. That is itself a web, independent of any groundwork or foundation [6] to support it. But it is possible to work it *over* a silken or other surface; and there is a kind of embroidery which only floats on the surface of the material without penetrating it. A fragment of last century silk given in Illustration 35 shows plainly what is meant.

larger image

3. STITCHING ON A SQUARE MESH.

Embroidery is enrichment by means of the needle. To embroider is to work *on* something: a groundwork is presupposed. And we usually understand by embroidery, needlework in thread (it may be wool, cotton, linen, silk, gold, no matter what) upon a textile material, no matter what. In short, it is the decoration of a material woven in thread by means still of thread. It is thus *the* consistent way of ornamenting stuff—most consistent of all when one kind of thread is employed throughout, as in the case of linen upon linen, silk upon silk. The enrichment may, however, rightly be, and oftenest is, perhaps, in a material nobler than the stuff enriched, in silk upon linen, in wool upon cotton, in gold upon velvet. The advisability of working upon a precious stuff in thread *less* precious is open to question. It does not seem to have been satisfactorily done; but if it were only the background that was worked, and the pattern were so schemed as almost to cover it, so that, in fact, very little of the more beautiful texture was sacrificed, and you had still a sumptuous pattern on a less attractive background—why not? But then it would be because you wanted that less [7] precious texture there. The excuse of economy would scarcely hold good.

In the case of a material in itself unsightly, the one course is to cover it entirely with stitching, as did the Persian and other untireable people of the East. But not they only. The famous Syon cope is so covered. Much of the work so done, all-over work that is to say, competes in effect with tapestry or other weaving; and its purpose was similar: it is a sort of amateur way of working your own stuff. But in character it is no more nearly related to the work of the loom than other needlework—it is still work *on* stuff. For all-over embroidery one chooses, naturally, a coarse canvas ground to work on; but it more often happens that one chooses canvas because one means to cover it, than that one works all over a ground because it is unpresentable.

Embroidery is merely an affair of stitching; and the first thing needful alike to the worker in it and the designer for it is, a thorough acquaintance with the stitches; not, of course, with every modification of a modification of a stitch which individual ingenuity may have devised—it would need the space of an encyclopædia to chronicle them all—but with the broadly marked varieties of stitch which have been employed to best purpose in ornament.

They are derived, naturally, from the stitches first used for quite practical and prosaic purposes—buttonhole [8] stitch, for example, to keep the edges of the stuff from fraying; herring-bone, to strengthen and disguise a seam; darning, to make good a worn surface; and so on.

The difficulty of discussing them is greatly increased by the haphazard way in which they are commonly named. A stitch is called Greek, Spanish, Mexican, or what not, according to the country whence came the work in which some one first found it. Each names it after his or her individual discovery, or calls it, perhaps, vaguely Oriental; and so we have any number of names for the same stitch, names which to different people stand often for quite different stitches.

When this confusion is complicated by the invention of a new name for every conceivable combination of thread-strokes, or for each slightest variation upon an old stitch, and even for a stitch worked from left to right instead of from right to left, or for a stitch worked rather longer than usual, the task of reducing them to order seems almost hopeless.

Nor do the quasi-learned descriptions of old stitches help us much. One reads about *opus* this and *opus* that, until one begins to wonder where, amidst all this parade of science, art comes in. But you have not far to go in the study of the authorities to discover that, though they may concur in using certain high-sounding Latin terms, they are not of the same mind as to their meaning. [9] In one thing they all agree, foreign writers as well as English, and that is, as to the difficulty of identifying the stitch referred to by ancient writers, themselves probably not acquainted with the *technique* of stitching, and as likely as not to call it by a wrong name. It is easier, for example, to talk of *Opus Anglicanum* than to say precisely what it was, further than that it described work done in England; and for that we have the simple word—English. There is nothing to show that mediæval English work contained stitches not used elsewhere. The stitches probably all come from the East.

Nomenclature, then, is a snare. Why not drop titles, and call stitches by the plainest and least mistakable names? It will be seen, if we reduce them to their native simplicity, that they fall into fairly-marked groups, or families, which can be discussed each under its own head.

Stitches may be grouped in all manner of arbitrary ways—according to their provenance, according to their effect, according to their use, and so on. The most natural way of grouping them is according to their structure; not with regard to whence they came, or what they do, but according to what they are, the way they are worked. This, at all events, is no arbitrary classification, and this is the plan it is proposed here to adopt. [10]

The use of such classification hardly needs pointing out.

A survey of the stitches is the necessary preliminary, either to the design or to the execution of needlework. How else suit the design to the stitch, the stitch to the design? In order to do the one the artist must be quite at home among the stitches; in order to do the other the embroidress must have sympathy enough with a design to choose the stitch or stitches which will best render it. An artist who thinks the working out of his sketch none of his business is no practical designer; the worker who thinks design a thing apart from her is only a worker.

This is not the moment to urge upon the needlewoman the study of design, but to urge upon the designer the study of stitches. Nothing is more impractical than to make a design without realising the labour involved in its execution. Any one not in sympathy with stitching may possibly design a beautiful piece of needlework, but no one will get all that is to be got out of the needle without knowing all about it. One must understand the ways in which work can be done in order to determine the way it shall in any particular case be done.

Certain stitches answer certain purposes, and strictly only those. The designer must know which stitch answers which purpose, or he will in the first place waste the labour of the embroidress, [11] and in the second miss his effect, which is to waste his own pains too. The effective worker (designer or embroiderer) is the one who works with judgment—and you cannot judge unless you know. When it is remembered that the character of needlework, and by rights also the character of its design, depends upon the stitch, there will be no occasion to insist further upon the necessity of a comprehensive survey of the stitches.

A stitch may be defined as the thread left on the surface of the cloth or what not, after each ply of the needle.

And the simple straightforward stitches of this kind are not so many as one might suppose. They may be reduced indeed to a comparatively few types, as will be seen in the following chapters.

CANVAS STITCHES. [12]

The simplest, as it is most likely the earliest used, stitch-group is what might best be called Canvas stitch — of which cross-stitch is perhaps the most familiar type, the class of stitches which come of following, as it is only natural to do, the mesh of a coarse canvas, net, or open web upon which the work is done.

A stitch bears always, or should bear, some relation to the material on which it is worked; but canvas or very coarse linen almost compels a stitch based upon the cross lines of its woof, and indeed suggests designs of equally rigid construction. That is so in embroidery no matter where. In ancient Byzantine or Coptic work, in modern Cretan work, and in peasant embroidery all the world over, pattern work on coarse linen has run persistently into angular lines — in which, because of that very angularity, the plain outcome of a way of working, we find artistic character. Artistic design is always expressive of its mode of workmanship.

Work of this kind is not too lightly to be dismissed. There is art in the rendering of form by means of angular outlines, art in the choice of [13] forms which can be expressed by such lines. It is not uncharitable to surmise that one reason why such work (once so universal and now quite out of fashion) is not popular with needlewomen may be, the demand it makes upon the designer's draughtmanship: it is much easier, for example, to draw a stag than to render the creature satisfactorily within jagged lines determined by a linen mesh.

larger image

4. CROSS-STITCH.

The piquancy about natural or other forms thus reduced to angularity argues, of course, no affectation of quaintness on the part of the worker, but was the unavoidable outcome of her way of work. There is a pronounced and early limit to art of this rather naïve kind, but that there is art in some of the very simplest and most modest peasant work built up on those lines no artist will deny. The art in it is usually in proportion to its modesty. Nothing is more futile than to put it to anything [14] like pictorial purpose. The wonderfully wrought pictures in tent-stitch, for example, bequeathed to us by the 17th century, are painful object lessons in what not to do.

The origin of the term cross-stitch is not far to seek: the stitches worked upon the square mesh do cross. But, falling naturally into the lines of the mesh which governs them, they present not so much the appearance of crosses as of squares, reminding one of the tesseræ employed in mosaic.

TO WORK CROSS STITCH.

To explain the process of working cross-stitch would be teaching one's grandmother indeed. It is simply, as its name implies, crossing one stitch by another, following always the lines of the canvas. But the important thing about it is that the stitches must cross always in the same way; and, more than that, they must be worked in the same direction, or the mere fact that the stitches at the *back* of the work do not run in the same way will disturb the evenness of the surface. What looks like a seam on the sampler opposite is the result of filling up a gap in the ground with stitches necessarily worked in vertical, whereas the ground generally is in horizontal, lines. On the face of the work the stitches cross all in the same way.

The common use of cross-stitch and the somewhat geometric kind of pattern to which it lends itself are shown in the sampler, Illustration 5.

The broad and simple leafage, worked solid (A) or left in the plain canvas upon a groundwork of [16] solid stitching (B), and the fretted diaper on vertical and horizontal lines (C), show the most straightforward ways of using it.

larger image

5. CROSS-STITCH SAMPLER.

The criss-cross of alternating cross-stitches and open canvas framed by the key pattern (C) shows a means of getting something like a tint halfway between solid work and plain ground. The mere work line — or "stroke-stitch," not crossed (D), is a perfectly fair way of getting a delicate effect; but the design has a way of working out rather less happily than it promised.

The addition of such stroke-stitches to solid cross-stitch (E) is not at best a very happy device. It strikes one always as a confession of dissatisfaction on the part of the worker with the simple means of her choice. As a device for, as it were, correcting the stepped outline it is at its worst. Timid workers are always afraid of the stepped outline which a coarse mesh gives. In that they are wrong. One should employ canvas stitch only where there is no objection to a line which keeps step with the canvas; then there is a positive charm (for frank people at least) in the frank confession of the way the work is done.

There are many degrees in the frankness with which this convention has been accepted, according perhaps to the coarseness of the canvas ground, perhaps to the personality of the worker. The animal forms at the top of Illustration 6 are uncompromisingly square; the floral devices on [18] the same page, though they fall, as it were inevitably, into square lines, are less rigidly formal. The inevitableness of the square line is apparent in the sprig below (7). It was evidently meant to be freely drawn, but the influence of the mesh betrays itself; and the design, if it loses something in grace, gains also thereby in character.

larger image
6. CANVAS-STITCH.

larger image

7. CANVAS-STITCH.

There is literally no end to the variety of stitches, as they are called, belonging to this group, and their names are a babel of confusion. Florentine, Parisian, Hungarian, Spanish, Moorish, Cashmere, Milanese, Gobelin, are only a few of them; but they stand, as a rule, rather for stitch arrangements than for stitches. A small selection of them is given in Illustration 8.

tent-stitch A.

What is known as tent-stitch (A in the sampler opposite) is a sort of half cross-stitch; its peculiarity is that it covers only one thread of the [20] canvas at a stroke, and is therefore on a more minute scale than stitches which are two or three threads wide, as cross-stitch may, and cushion-stitch must, be. It derives its name from the old word tenture, or tenter (*tendere*, to stretch), the frame on which the embroidress distended her canvas. The word has gone out of use, but we still speak of tenter-hooks. The stitch is serviceable enough in its way, but is discredited by the monstrous abuse of it referred to already. A picture in tent-stitch is even more foolish than a picture in mosaic. It cannot come anywhere near to pictorial effect; the tesseræ will pronounce themselves, and spoil it.

larger image
8. CANVAS-STITCH SAMPLER.

larger image

9. CUSHION AND SATIN STITCHES.

cross-stitch B.

This kind of half cross-stitch worked on the larger scale of ordinary cross-stitch would look [21] meagre. It is filled out, therefore (B), by horizontal lines of the thread laid across the canvas, and over these the stitch is worked.

cushion-stitch C.

Cushion-stitch consists of diagonal lines of upright stitches, measuring in the sampler (C) six threads of the canvas, so that after each stitch the needle may be brought out just three threads lower than where it was put in. By working in zigzag instead of diagonal lines, a familiar pattern is produced, more often described as "Florentine;" but the stitch is in any case the same.

canvas-stitch D.

The stitch at D (sometimes called Moorish stitch) is begun by working a row of short vertical stitches, slightly apart, and completed by diagonal stitches joining them.

Unless the silk employed is full and soft, this may not completely cover the canvas, in which case the diagonal stitches must further be crossed as shown on Illustration 89.

If the linen is loosely woven and the thread is tightly drawn in the working, the mesh is pulled apart, giving the effect of an open lattice of the kind shown at B, on Illustration 10, in which the threads of the linen are not drawn out but drawn together.

canvas-stitch E.

The way of working the stitch at E is described on page 51, under the name of "fish-bone." Worked on canvas it has somewhat the effect of plaiting, and goes by the name of "plait-stitch." [22] It is worked in horizontal rows alternately from left to right and from right to left.

canvas-stitch F.

The stitch at F is a sort of couching (see page 124). Diagonal lines of thread are first laid from edge to edge of the ground space, and these are sewn down by short overcasting stitches in the cross direction.

Admirable canvas stitch work has been done upon linen in silk of one colour—red, green, or blue—and it was a common practice to work the background leaving the pattern in the bare stuff. It prevailed in countries lying far apart, though probably not without inter-communication. In fact, the influence of Oriental work upon European has been so great that even experts hesitate sometimes to say whether a particular piece of work is Turkish or Italian. In Italian work, at least, it was usual to get over the angularity of silhouette inherent in canvas stitches by working an outline separately. When that is thin, the effect is proportionately feeble. The broader outline (shown at A, Illustration 10) justifies itself, and in the case of a stitch which falls into horizontal lines, it appears to be necessary. This is plait stitch, known also by the name of Spanish stitch—not that it is in any way peculiar to Spain. It is allied to herring-bone-stitch, to which a special chapter is devoted.

10. PLAIT AND OPEN CANVAS STITCHES.

Darning is also employed as a canvas stitch. There is beautiful 16th century Italian work (in [24] coloured silks on dark net of the very open square mesh of the period), which is most effective, and in which there is no pretence of disguising the stepped outline; and in the very early days of Christian art in Egypt and Byzantium, linen was darned in little square tufts of wool upstanding on its surface, which look so much like the tesseræ of mosaic that it seems as if they must have been worked in deliberate imitation of it.

Again, in the 15th century satin-stitch was worked on fine linen with strict regard to the lines of its web; and the Persians, ancient and modern, embroider white silk upon linen, also in satin-stitch, preserving piously the rectangular and diagonal lines given by the material. They have their reward in producing most characteristic needlework. The diapered ground in Illustration 9 (page 20) is satin-stitch upon coarse linen.

The filling-in patterns used to such delicate and dainty purpose in the marvellous work on fine cambric (Illustration 73) which competes in effect with lace, though it is strictly embroidery, all follow in their design the lines of the fabric, and are worked thread by thread according to its woof: they afford again instances of perfect adaptation of stitch to material and of design to stitch.

Satin and other stitches were worked by the old Italians (Illustration 3) on square-meshed canvas, frankly on the square lines given by it, for the filling in of ornamental details, though the [25] outline might be much less formal. That is to say, the surface of freely-drawn leaves, &c., instead of being worked solid, was diapered over with more or less open pattern work constructed on the lines of the weaving.

A cunning use of the square mesh of canvas has sometimes been made to guide the worker upon other fabrics, such as velvet. This was first faced with net: the design was then worked, over that, on to and into the velvet, and the threads of the canvas were then drawn out. That is a device which may serve on occasion. The design may even be traced upon the net.

CREWEL-STITCH. [26]

For work in the hand, Crewel-Stitch is perhaps, on the whole, the easiest and most useful of stitches; whence it comes that people sometimes vaguely call all embroidery crewel work; though, as a matter of fact, the stitch properly so called was never very commonly employed, even when the work was done in "crewel," the double thread of twisted wool from which it takes its name.

the working of A on crewel-stitch sampler.

larger image
11. CREWEL-STITCH SAMPLER.

larger image

12. CREWEL-STITCH SAMPLER (BACK).

to work A.

Crewel-Stitch proper is shown at A on the sampler opposite, where it is used for line work. It is worked as follows:—Having made a start in the usual way, keep your thread downwards under your left thumb and below your needle—that is, [29] to the right; then take up with the needle, say ⅛th of an inch of the stuff, and bring it out through the hole made in starting the stitch, taking care not to pierce the thread. This gives the first half stitch. If you proceed in the same way your next stitch will be full length. The test of good workmanship is that at the back it should look like back-stitch (Illustration 12), described on page 30.

the working of B on crewel-stitch sampler.

to work B.

Outline-Stitch (B on sampler) differs from crewel-stitch only in that the thread is always kept upwards above the needle, that is to the left. In so doing the thread is apt to untwist itself, and wants constantly re-twisting. The stitch is useful for single lines and for outlining solid work. The muddled effect of much crewel work is due to the confusion of this stitch with crewel-stitch proper.

to work C.

Thick Crewel-Stitch (C on sampler) is only a little wider than ordinary crewel-stitch, but gives a heavier line, in higher relief. In effect it resembles rope-stitch, but it is more simply worked. You begin as in ordinary crewel-stitch, but after the first half-stitch you take up ⅛th of [30] an inch of the material in advance of the last stitch, and bring out your needle at the point where the first half-stitch began. You proceed, always putting your needle in ⅛th of an inch in front of, and bringing it out ⅛th of an inch behind, the last stitch, so as to have always ¼th of an inch of the stuff on your needle.

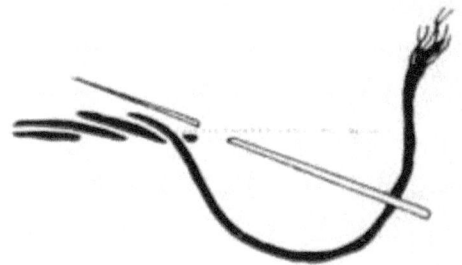

the working of G on crewel-stitch sampler.

to work D.

Thick Outline-Stitch (D on sampler) is like thick crewel-stitch with the exception that, as in ordinary outline-stitch (B), you keep your thread always above the needle to the left.

to work E.

In Back-Stitch (E), instead of first bringing the needle out at the point where the embroidery is to begin, you bring it out ⅛th of an inch in advance of it. Then, putting your needle back, you take up this ⅛th together with another ⅛th in advance. For the next stitch you put your needle into the hole made by the last stitch, and so on, taking care not to split the last thread in so doing.

to work F.

To work the Spots (F) on sampler—having made a back-stitch, bring your needle out through the same hole as before, and make another back-stitch above it, so that you have, in what appears to be one stitch, two thicknesses of thread; then [32] bring your needle out some distance in advance of the last stitch, and proceed as before. The distance between the stitches is determined by the effect you desire to produce. The thread should not be drawn too tight.

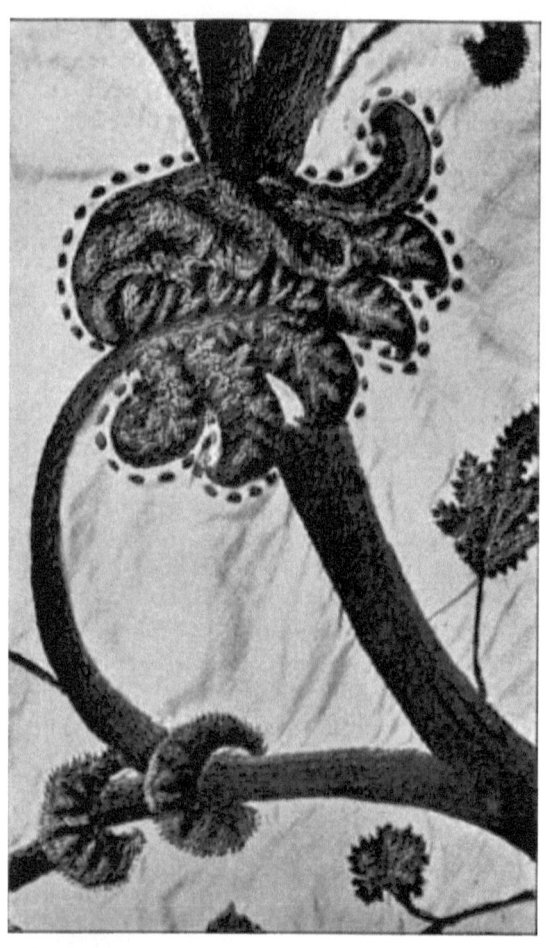

larger image

13. CREWEL WORK AND CREWEL-STITCH.

to work G.

You begin Stem-Stitch (G) with the usual half-stitch. Then, holding the thread downwards, instead of proceeding as in crewel-stitch (A) you slant your needle so as to bring it out a thread or two higher up than the half-stitch, but precisely above it. You next put the needle in ⅛th of an inch in advance of the last stitch, and, as before, bring it out again in a slanting direction a thread or two higher. At the back of the work (Illustration 12) the stitches lie in a slanting direction.

to work H.

To work wider Stem-Stitch (H). After the first two stitches, bring your needle out precisely above and in a line with them, and put it in again 1/8th of an inch in advance of the last stitch, producing a longer stroke, which gives the measure of those following. The slanting stitches at the back (Illustration 12) are only two-thirds of the length of those on the face.

Crewel and Outline Stitches worked (J) side by side give somewhat the effect of a braid. The importance of not confusing them, already referred to, is here apparent.

Crewel-Stitch is worked SOLID in the heart-shape in the centre of the sampler. On the left side the rows of stitching follow the outline of [34] the heart; on the right they are more upright, merely conforming a little to the shape to be filled. This is the better method.

larger image

14. CREWEL WORK IN VARIOUS STITCHES.

TO WORK SOLID CREWEL-STITCH.

The way to work solid crewel-stitch will be best explained by an instance. Suppose a leaf to be worked. You begin by outlining it; if it is a wide leaf, you further work a centre line where the main rib would be, and then work row within row of stitches until the space is filled. If on arriving at the point of your leaf, instead of going round the edge, you work back by the side of the first row of stitching, there results a streakiness of texture, apparent in the stem on Illustration 13. What you get is, in effect, a combination of crewel and outline stitches, as at J, which in the other case only occurs in the centre of the shape where the files of stitches meet.

To represent shading in crewel-stitch, to which it is admirably suited (A, Illustration 41), it is well to work from the darkest shadows to the highest lights. And it is expedient to map out on the stuff the outline of the space to be covered by each shade of thread. There is no difficulty then in working round that shape, as above explained.

In solid crewel the stitches should quite cover the ground without pressing too closely one against the other.

larger image

15. CREWEL-STITCH IN TWISTED SILK.

It does not seem that Englishwomen of the 17th century were ever very faithful to the stitch we know by the name of crewel. Old examples of [36] work done entirely in crewel-stitch, as distinguished from what is called crewel work, are seldom if ever to be met with. The stitch occurs in most of the old English embroidery in wool; but it is astonishing, when one comes to examine the quilts and curtains of a couple of hundred years or so ago, how very little of the woolwork on them is in crewel-stitch. The detail on Illustration 13 was chosen because it contained more of it than any other equal portion of a handsome and typical English hanging; but it is only in the main stem, and in some of the outlines, that the stitch is used. And that appears to have been the prevailing practice—to use crewel-stitch for stems and outlines, and for little else but the very simplest forms. The filling in of the leafage, the diapering within the leaf shapes, and the smaller and more elaborate details generally were done in long-and-short-stitch, or whatever came handiest. In fact, the thing to be represented, fruit, berry, flower, or what not, seems to have suggested the stitch, which it must be confessed was sometimes only a sort of scramble to get an effect.

Of course the artist always chooses her stitch, and she is free to alter it as occasion may demand; but a good workwoman (and the embroidress is a needlewoman first and an artist afterwards, perhaps) adopts in every case a method, and departs from it only for very good reason. It looks as if our ancestors had set to work without system or guiding principle at all. No doubt they [37] got a bold and striking effect in their bed-hangings and the like; but there is in their work a lack of that conscious aim which goes to make art. Theirs is art of the rather artless sort which is just now so popular. Happily it was kept in the way it should go by a strict adherence to traditional pattern, which for the time being seems to have gone completely out of fashion.

Quite in the traditional manner is Illustration 14. One would fancy at first sight that the work was almost entirely in crewel-stitch. As a matter of fact, there is little which answers to the name, as an examination of the back of the work shows plainly enough. What the stitches are it is not easy to say. The mystery of many a stitch is to be unravelled only by literally picking out the threads, which one is not always at liberty to do, although, in the ardour of research, a keen embroidress will do it—not without remorse in the case of beautiful work, but relentlessly all the same.

The only piece of embroidery entirely in crewel-stitch which I could find for illustration (15) is worked, as it happens, in silk; nor was the worker aware that in so working she was doing anything out of the common. Another instance of crewel-stitch is given in the divided skirt, let us call it, of the personage in Illustration 72.

Beautiful back-stitching occurs in the Italian work on Illustration 89, and the stitch is used for sewing down the *appliqué* in Illustration 94.

CHAIN-STITCH. [38]

larger image

16. CHAIN-STITCH AND KNOTS.

Chain and Tambour Stitch are in effect practically the same, and present the same rather granular surface. The difference between them is that chain-stitch is done in the hand with an ordinary needle, and tambour-stitch in a frame with a hook sharper at the turning point than an ordinary crochet hook. One takes it rather for granted that work which was presumably done in the hand (a large quilt, for example) is chain-stitch, and that what seems to have been done in a frame is tambour work, though it is possible, but not advisable of course, to work chain-stitch in a frame.

Chain-stitch is not to be confounded with split-stitch (see page 105), which somewhat resembles it.

to work A.

To work chain-stitch (A on the sampler, Illustration 17) bring the needle out, hold the thread [41] down with the left thumb, put the needle in again at the hole through which you brought it out, take up ¼ of an inch of stuff, and draw the thread through: that gives you the first link of the chain. The back of the work (18) looks like back-stitch. In fact, in the quilted coverlet, Illustration 69 (as in much similar work of the period), the outline pattern, which you might take for back-stitching, proves to have been worked from the back in chain-stitch. The same thing occurs in the case of the Persian quilt in Illustration 70.

17. CHAIN-STITCH SAMPLER.

larger image

18. CHAIN-STITCH SAMPLER (BACK).

to work B.

A playful variation upon chain-stitch (B on the sampler, Illustration 17) is effected by the use of two threads of different colour. Take in your needle a dark and a light thread, say the dark one to the left, and bring them out at the point at which your work begins. Hold the dark thread under your thumb, and, keeping the light one to the right, well out of the way, draw both threads through; this makes a dark link; the light thread disappears, and comes out again to the left of the dark one, ready to be held under the thumb while you make a light link. This "magic stitch," as it has been called, is no new invention. It is to be found in Persian, Indian, and Italian Renaissance work. An instance of it occurs in Illustration 64.

to work C.

A variety of chain-stitch (C on the sampler, Illustration 17) used often in church work, more solid in appearance, the links not being so open, [42] is rather differently done. Begin a little in advance of the starting point of your work, hold the thread under your thumb, put the needle in again at the starting point slightly to the left, bring your needle out about ⅛th of an inch below where it first went in but precisely on the same line, and you have the first link of your chain.

to work D.

To work what is known as cable-chain (D on the sampler, Illustration 17) keep your thread to the right, put in your needle, pointing downwards, a little below the starting point, and bring it out about ¼th of an inch below where you put it in; then put it through the little stitch just formed, from right to left, hold your thread towards the left under your thumb, put your needle through the stitch now in process of making from right to left, draw up the thread, and the first two links of your chain are made.

to work E.

A zigzag chain, of a rather fancy description, goes by the name of Vandyke chain (E on the sampler, Illustration 17). To make it, bring your needle out at a point which is to be the left edge of your work, and make a slanting chain-stitch from left to right; then, putting your needle into that, make another slanting stitch, this time from right to left—and so to and fro to the end.

to work F.

The braid-stitch shown at F on the sampler (Illustration 17) is worked as follows, horizontally from right to left. Bring your needle out at a point which is to be the lower edge of your work, [43] throw your thread round to the left, and, keeping it all the time loosely under your thumb, put your needle under the thread and twist it once round to the right. Then, at the upper edge of your work, put in the needle and slide the thread towards the right, bring the needle out exactly below where you put it in, carry your thread under the needle towards the left, draw the thread tight, and your first stitch is done.

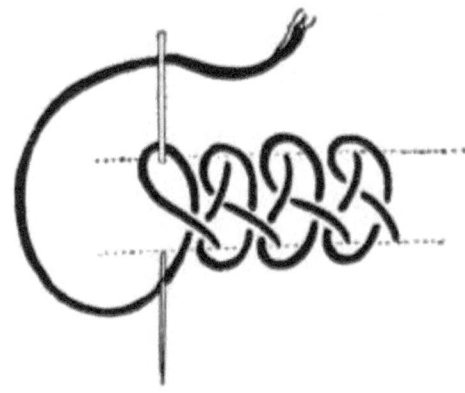

the working of F on chain-stitch sampler.

to work G.

A yet more fanciful variety of braid-stitch (G on the sampler, Illustration 17) is worked vertically, downwards. Having, as before, put your needle under the thread and twisted it once round, put it in at a point which is to be the left edge of your work, and, instead of bringing it out immediately below that point, slant it to the right, bringing it out on that edge of the work, and finish your stitch as in the case of F.

These braid-stitches look best worked in stout thread of close texture.

In covering a surface with chain-stitch (needlework or tambour) the usual plan is to follow the [44] contour of the design, working chain within chain until the leaf or whatever it may be is filled in. This stitch is rarely worked in lines across the forms, but it has been effectively used in that way, following always the lines of the warp and weft of the stuff. Even in that case the successive lines of stitching should be all in one direction—not running backwards and forwards—or it will result in a sort of pattern of braided lines. The reason for the more usual practice of following the outline of the design is obvious. The stitch lends itself to sweeping, even to perfectly spiral, lines—such as occur in Greek wave patterns: it was, in fact, made use of in that way by the Greeks some four or five centuries B.C.

the working of G on chain-stitch sampler.

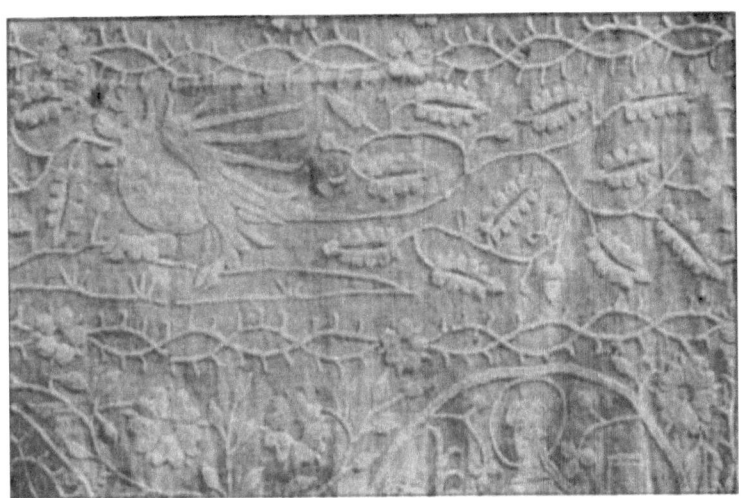

larger image

19. CHAIN AND SURFACE STITCHES.

We owe the tambour frame, they say, to China; but it has been largely used, and abused indeed, in England. Tambour work, when once you have the trick of it, is very quickly done—in about one-sixth of the time it would take to do it with the needle. It has the further advantage that it serves equally well for embroidery on a light or on a heavy stuff, and that it is most lasting. The misfortune is that the sewing [46] machine has learnt to do something at once so like it and so mechanically even, as to discredit genuine handwork, whether tambour work or chain-stitch. For all that, neither is to be despised. If they have often a mechanical appearance that is not all the fault of the stitch: the worker is to blame. Indian embroiderers depart sometimes so far from mechanical precision as to shock the admirers of monotonously even work. Artistic use of chain stitch is made in many of our illustrations: for outlines in Illustrations 24 and 72; for surface covering in Mr. Crane's lion, Illustration 74; to represent landscape in Illustration 78, where everything except the faces of the little men is in chain-stitch; and again for figure work in Illustration 81. In Illustration 19 it occurs in association with a curious surface stitch; in Illustration 64 it is used to outline and otherwise supplement inlay. The old Italians did not disdain to use it. In fact, wherever artists have employed it, they show that there is nothing inherently inartistic about the stitch.

HERRING-BONE STITCH. [47]

Herring-bone is the name by which it is customary to distinguish a variety of stitches somewhat resembling the spine of a fish such as the herring. It would be simpler to describe them as "fish-bone;" but that term has been appropriated to describe a particular variety of it. One would have thought it more convenient to use fish for the generic term, and a particular fish for the specific. However, it saves confusion to use names as far as possible in their accepted sense.

It will be seen from the sampler, Illustration 20, that this stitch may be worked open or tolerably close; but in the latter case it loses something of its distinctive character. Fine lines may be worked in it, but it appears most suited to the working of broadish bands and other more or less even-sided or, it may be, tapering forms, more feathery in effect than fish-bone-like, such as are shown at E on sampler.

Ordinary herring-bone is such a familiar stitch that the necessity of describing it is rather a matter of literary consistency than of practical importance. [48]

The two simpler forms of herring-bone (it is always worked from left to right, and begun with a half-stitch) marked A and C on the sampler are strikingly different in appearance, and are worked in different ways—as will be seen at once by reference to the back of the sampler (Illustration 21), where the stitches take in the one case a horizontal and in the other a vertical direction.

to work A.

To work A, bring your needle out about the centre of the line to be worked; put it into the lower edge of the line about ⅛th of an inch further on; take up this much of the stuff, and, keeping the thread to the right, above the needle, draw it through. Then, with the thread below it, to the right, put your needle into the upper edge of the line ¼th of an inch further on, and, turning it backwards, take up again ⅛th of an inch of stuff, bringing it out immediately above where it went in on the lower edge.

to work B.

What is called "Indian Herring-bone" (B) is merely stitch A worked in longer and more slanting stitches, so that there is room between them for a second row in another colour, the two colours being, of course, properly interlaced.

to work C.

To work C, bring your needle out as for A, and, putting it in at the upper edge of the line to be worked and pointing it downwards, whilst your thread lies to the right, take up ever so small a piece of the stuff. Then, slightly in advance of the last stitch, the thread still to the right, [51] your needle now pointing upwards, take another similar stitch from the lower edge.

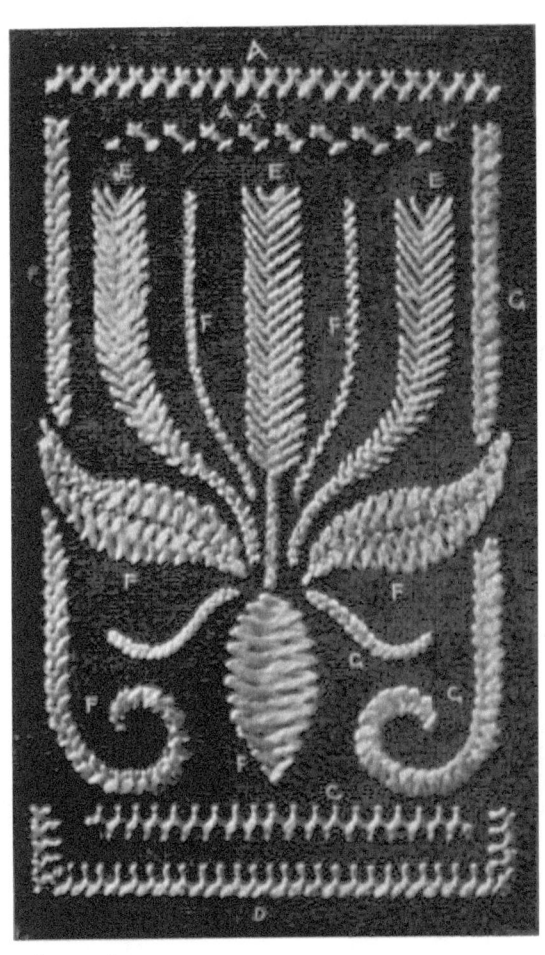

larger image

20. HERRING-BONE SAMPLER.

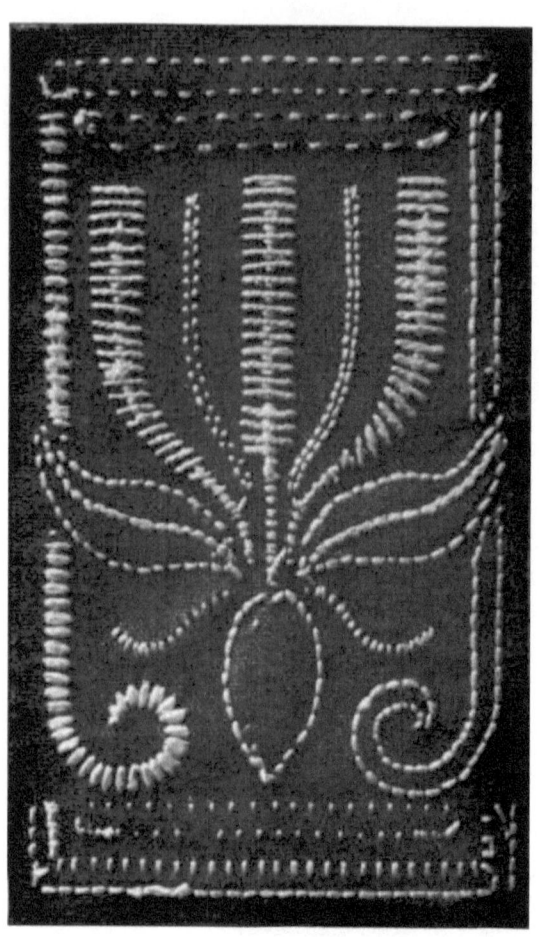

21. HERRING-BONE SAMPLER (BACK).

to work D.

The variety at D is merely a combination of A and C, as may be seen by reference to the back of the sampler (opposite); though the short horizontal stitches there seen meet, instead of being wide apart as in the case of A.

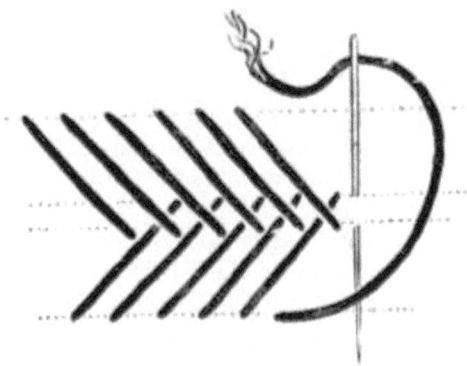

the working of E on herring-bone sampler.

to work E.

What is known as "fish-bone" is illustrated in the three feathery shapes on the sampler (E), two of which are worked rather open. It is characteristic of this stitch that it has a sort of spine up the centre where the threads cross. Suppose the stitch to be worked horizontally. Bring your needle out on the under edge of the spine about ¼th of an inch from the starting point of the work, and put it in on the upper edge of the work at the starting point, bringing it out immediately below that on the lower edge of the work. Put it in again on the upper edge of the spine, rather in [52] advance of where it came out on the lower edge of it before, and bring it out on the lower edge of this spine immediately below where it entered.

the working of F on herring-bone sampler.

to work F.

In close herring-bone (F on the sampler, Illustration 20) you have always a long stitch from left to right, crossed by a shorter stitch which goes from right to left. Having made a half stitch, bring the needle out at the beginning of the line to be worked, at the lower edge, and put it in ⅛th of an inch from the beginning of the upper edge. Bring it out again at the beginning of this edge and put it in at the lower edge ¼th of an inch from the beginning, bringing it out on the same edge ⅛th of an inch from the beginning. Put the needle in again on the upper edge ⅛th of an inch in front of the last stitch on that edge, and bring it out again, without splitting the thread, on the same edge as the hole where the last stitch went in.

If you wish to cover a surface with herring-bone-stitch, you work it, of course, close, so that each [53] successive stitch touches its foregoer at the point where the needle enters the stuff (F on the sampler, Illustration 20). It will be seen that at the back (21) this looks like a double row of back-stitching. Worked straight across a wide leaf, as in the lower half of sampler, it is naturally very loose. A better method of working is shown in the side leaves, which are worked in two halves, beginning at the base of a leaf on one side and working down to it on the other. There is here just the suggestion of a mid-rib between the two rows.

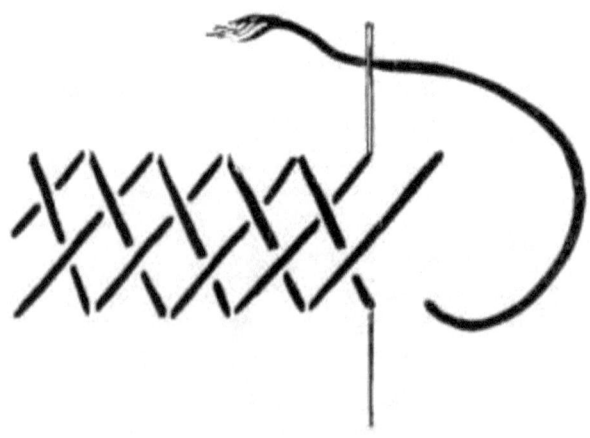

the working of G on herring-bone sampler.

to work G.

The stitch at G on sampler, having the effect of higher relief than ordinary close herring-bone (F), is sometimes misleadingly described as tapestry stitch. It is worked, as the back of the sampler (21) clearly shows, in quite a different way. You get there parallel rows of double stitches. Having [54] made a half-stitch entering the material at the upper edge of the work, bring the needle out on the lower edge of it immediately opposite. Then, going back, put it in at the beginning of the upper edge, and bring it out at the beginning of the lower one. Thence take a long slanting stitch upwards from left to right, bring the needle out on the lower edge immediately opposite, cross it by a rather shorter stitch from right to left, entering the stuff at the point where the first half-stitch ended, bring this out on the lower edge, opposite, and the stitch is done.

The artistic use of herring-bone-stitch is shown in the leaves of the tulip (84), and a closer variety of it in the pink, or whatever the flower may be, in the hand of the little figure on Illustration 72.

BUTTONHOLE-STITCH. [55]

Buttonhole is more useful in ornament than one might expect a stitch with such a very utilitarian name to be. It is, as its common use would lead one to suppose, pre-eminently a one-edged stitch, a stitch with which to mark emphatically the outside edge of a form. There is, however, a two-edged variety known as ladder-stitch, shown in the two horn shapes on the sampler, Illustration 22.

By the use of two rows back to back, leaf forms may be fairly expressed. In the leaves on the sampler, the edge of the stitch is used to emphasise the mid rib, leaving a serrated edge to the leaves. The character of the stitch would have been better preserved by working the other way about, and marking the edge of the leaves by a clearcut line, as in the case of the solid leaves in Illustration 73.

The stitch may be used for covering a ground or other broad surface, as in the pot shape (J) on the sampler, where the diaper pattern produced by its means explains itself the better for being worked in two shades of colour.

The simpler forms of the stitch are the more [56] useful. Worked in the form of a wheel, as in the rosettes at the side of the vase shape (A), the ornamental use of the stitch is obvious.

to work A.

One need hardly describe Buttonhole Stitch. The simple form of it (A) is worked by (when you have brought your needle out) keeping the thread under your thumb to the right, whilst you put the needle in again at a higher point slightly to the right, and bring it out immediately below, close to where it came out before. This and other one-edged stitches of the kind are sometimes called "blanket-stitch."

The only difference between versions such as B and C on the sampler, and simple buttonhole, is that the stitches vary in length according to the worker's fancy.

to work E.

The Crossed Buttonhole Stitch at E is worked by first making a stitch sloping to the right, and then a smaller buttonhole-stitch across this from the left.

The border marked D in sampler consists merely of two rows of slanting buttonhole-stitch worked one into the other. Needlewomen have wilful ways of making what should be upright stitches slant awkwardly in all manner of ways, with the result that they look as if they had been pulled out of the straight.

larger image

22. BUTTONHOLE SAMPLER.

larger image

23. BUTTONHOLE SAMPLER (BACK).

to work F.

The border at F, known as "Tailor's Buttonhole," is worked with the firm edge from you, instead of towards you, as you work ordinary [59] buttonhole. Bringing the thread out at the upper edge of the work to the left, and letting it lie on that side, you put your needle in again still on the same edge, and bring it out, immediately below, on the lower one. You then, before drawing the thread quite through, put your needle into the loop from behind, and tighten it upwards.

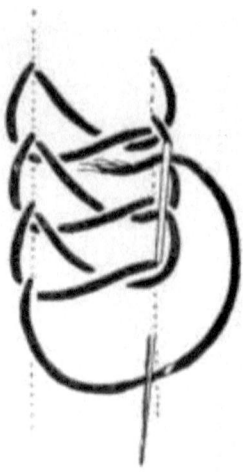

the working of H on buttonhole sampler.

to work G.

In order to make your ladder-stitch (G) square at the end, you begin by making a bar of the width the stitch is to be. Then, holding the thread under your thumb to the right, you put the needle in at the top of the bar and, slanting it towards the right, bring it out on a level with the other end of the bar somewhat to the right. This makes a triangle. With the point of your needle, pull the slanting thread out at the top, to form a square; insert the needle; slant it again to the right; draw it out as before, and you have your second triangle.

to work H.

The difference between the working of the lattice-like band at H, and ladder-stitch G, is that, having completed your first triangle, you make, by buttonholing a stitch, a second triangle pointing the other way, which completes a rectangular shape.

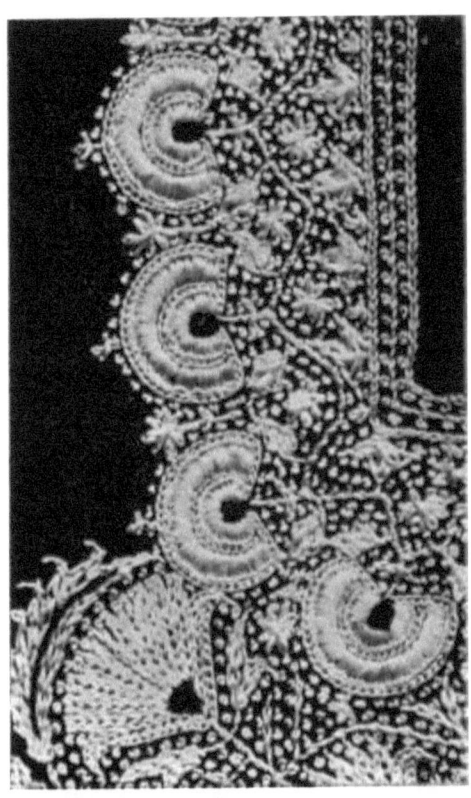

larger image

24. BUTTONHOLE, CHAIN, AND KNOT STITCHES.

In the solid work shown at J, you make five [61] buttonhole-stitches, gathering them to a point at the base, then another five, and so on. Repeat the process, this time point upwards, and you have the first band of the pot shape.

Characteristic and most beautiful use is made of buttonhole stitch in the piece of Indian work in Illustration 24, where it is outlined with chain stitch, which goes most perfectly with it.

Cut work, such as that on Illustration 65, is strengthened by outlining it in buttonhole-stitch.

Ladder-stitch occurs in the cusped shapes framing certain flowers in Illustration 72, embroidered all in blue silk on linen. It is not infrequent in Oriental work, and, in fact, goes sometimes by the name of Cretan-stitch on that account.

FEATHER AND ORIENTAL STITCHES. [62]

Feather-stitch is simply buttonholing in a slanting direction, first to the right side and then to the left, keeping the needle strokes in the centre closer together or farther apart according to the effect to be produced.

It owes its name, of course, to the more or less feathery effect resulting from its rather open character. Like buttonhole, it may be worked solid, as in the leaf and petal forms on the sampler, Illustration 25, but it is better suited to cover narrow than broad surfaces. The jagged outline which it gives makes it useful in embroidering plumage, but it is not to be confounded with what is called "plumage-stitch," which is not feather-stitch at all, but a version of satin-stitch.

The feathery stem (A) on the sampler is simply a buttonholing worked alternately from right to left and left to right.

to work B.

The border line at B requires rather more explanation. Presume it to be worked vertically. Bring your needle out at the left edge of the band; put it in at the right edge immediately opposite, keeping your thread under the needle to the right; [65] bring it out again still on the right edge a little lower down, and then, keeping your thread to the left, put the needle in on the left edge, opposite to where you last brought it out, and bring it out again on the same edge a little lower down.

25. FEATHER-STITCH SAMPLER.

larger image

26. FEATHER-STITCH SAMPLER (BACK).

The border at C is merely an elaboration of the above, with three slanting stitches on each edge instead of a single one in the direction of the band.

the working of G G on feather-stitch sampler.

Bands D, E, F, G, are variations of ordinary feather-stitch, requiring no further explanation than the back view of the work (26) affords. On the face of the sampler it will be noticed that lines have been drawn for the guidance of the worker. These are always four in number, indicating at once, that the stitch is made with four strokes of the needle, and the points at which it is put in and out of the stuff.

to work G G.

In working G G, suppose four guiding lines to have been drawn as above—numbered, 1, 2, 3, 4, from left to right. Bring your needle out at the top of line 1. Make a chain-stitch slanting downwards from line 1 to line 2. Put your needle into line 3 about ⅛th of an inch lower down, and, slanting it upwards, [66] bring it out on line 4 level with the point where you last brought it out. Make a chain-stitch slanting downwards this time from right to left, and bring your needle out on line 3. Lastly, put your needle into line 2, ⅛th of an inch below the last stitch, and, slanting it upwards, bring it out on line 1.

Feather-stitch is not adapted to covering broad surfaces solidly, but may be used for narrow ones.

Oriental-stitch is the name given to a close kind of feather-stitch much used in Eastern work. The difference at once apparent to the eye between the two is that, whereas for the mid-rib of a band or leaf of feather-stitching (25) you have cross lines, in Oriental-stitch (27) you have a straight line—longer or shorter as the case may be.

Oriental-stitch, sometimes called "Antique-stitch," is a stitch in three strokes, just as feather-stitch is a stitch in four. It is usually worked horizontally, though shown upright on the sampler, Illustration 27. Like feather-stitch (see diagram), it is worked on four guiding lines, faintly visible on the sampler.

to work A, B, C.

Stitches A, B, and C are worked in precisely the same way. Bring your needle out at the top of line 1. Keep the thread under your thumb to the right and put your needle in at the top of line 4, bringing it out into line 3 on the same level. Then put it in again at line 2, just on the other side of the thread, and bring it out on line 1 ready to begin the next stitch.

larger image
27. ORIENTAL-STITCH SAMPLER.

larger image

28. ORIENTAL-STITCH SAMPLER (BACK).

[69] It will be seen that the length of the central part (or mid-rib, as it was called above) makes the whole difference between the three varieties of stitch. In A the three parts are equal: in B the mid-rib is narrow: in C it is broad, as is most plainly seen on the back of the sampler (28). The difference is only a difference of proportion.

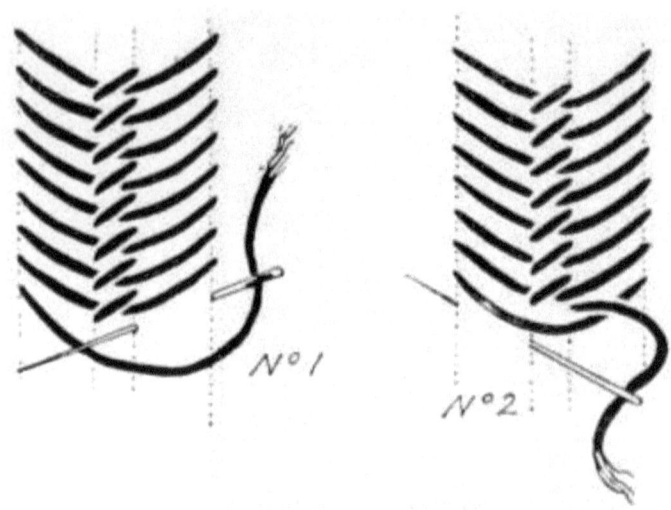

the working of A, B, C on oriental-stitch sampler.

to work D.

The sloping stitch at D is worked in the same way as A, B, C, except that instead of straight strokes with the needle you make slanting ones.

to work E.

Stitch E differs from D in that the side strokes slant both in the same direction. It is worked from right to left instead of from left to right.

to work F.

Stitch F is a combination of buttonhole and Oriental stitches. Between two rows of buttonholing [70] (dark on sampler) a single row of Oriental-stitch is worked.

The stitch employed for the central stalk, G, has really no business on this sampler, except that it has something of the appearance of a continuous Oriental-stitch.

Oriental-stitch is one of the stitches used in Illustration 72.

ROPE AND KNOT STITCHES. [71]

A single sampler is devoted to Rope and Knotted Stitches, more nearly akin than they look, for rope-stitch is all but knotted as it is worked.

Rope-stitch is so called because of its appearance. It takes a large amount of silk or wool to work it, but the effect is correspondingly rich. It is worked from right to left, and is easier to work in curved lines than in straight.

to work A, B.

Lines A on the sampler, Illustration 29, represent the ordinary appearance of the stitch; its construction is more apparent in the central stalk B, which is a less usual form of the same stitch, worked wider apart.

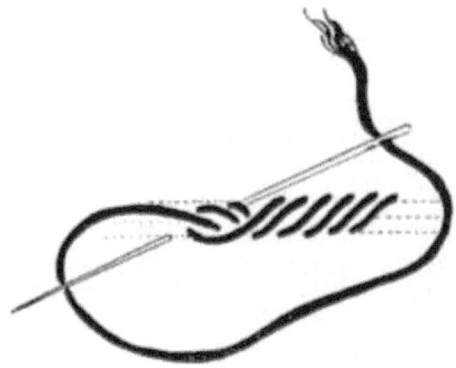

the working of A, B, on rope-stitch sampler.

Having brought out your needle at the right end of the work, hold part of the thread towards the left, under the thumb, the rest of it falling to the right; put your [72] needle in above where it came out, slant it towards you, and bring it out again a little in advance of where it came out before, and just below the thread held under your thumb. Draw the thread through, and there results a stitch which looks rather like a distorted chain stitch (B). The next step is to make another similar stitch so close to the foregoing one that it overlaps it partly. It is this overlapping which gives the stitch the raised and rope-like appearance seen at A.

the working of C on rope-stitch sampler.

to work C.

A knotted line (C in the sampler, Illustration 29) is produced by what is known as "German Knot-stitch," effective only in thick soft silk or wool. Begin as in rope stitch, keeping your thread in the same position. Then put your needle into the stuff just above the thread stretched under your thumb, and bring it out just below and in a line with where it went in; lastly, keep the needle above the loose end of the thread, draw it through, tightening the thread upwards, and you have the first of your knots: the rest follow at intervals determined by your wants.

to work D.

The more open stitch at D is practically the same [75] thing, except that in crossing the running thread you take up more of the stuff on each side of it.

larger image

29. ROPE-STITCH AND KNOT-STITCH SAMPLER.

larger image

30. ROPE-STITCH AND KNOT-STITCH SAMPLER (BACK).

to work E.

What is known by the name of "Old English Knot-stitch" (E) is a much more complicated stitch. Keeping your thread well out of the way to the right, put your needle in to the left, and take up vertically a piece of the stuff the width of the line to be worked at its widest, and draw the thread through. Then, keeping it under the thumb to the left, put your needle, eye first, downwards, through the slanting stitch just made; draw the thread not too tight, and, keeping it as before under the thumb, put your needle, eye first, this time through the upper half only of the slanting stitch, making a kind of buttonhole-stitch round the last, and draw out your thread.

These knotted rope stitches, call them what you will, are rather ragged and fussy—not much more than fancy stitches—of no great importance. Knots used separately are of much more artistic account.

to work F.

Bullion or Roll-stitch is shown in its simplest form in the petals of the flowers F on the sampler, Illustration 29. To work one such petal, begin by attaching the thread very firmly; bring your needle out at the base of the petal, put it in at the tip, and bring it out once more at the base, only drawing it partly through. With your right hand wind the thread, say seven times, round the projecting point of the needle from left to right. Then, [76] holding the coils under your left thumb, your thread to the right, draw your needle and thread through; and, dropping the needle, and catching the thread round your little finger, take hold of the thread with your thumb and first finger and draw the coiled stitch to the right, tightening it gently until quite firm. Lastly, put the needle through at the tip of the petal, and the stitch is complete and ready to be fastened off.

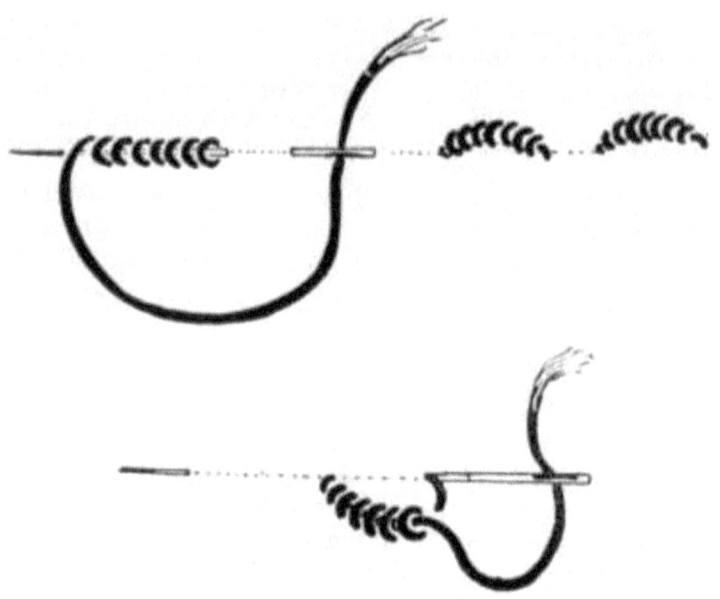

the working of F on knot-stitch sampler.

The leaves of these flowers consist simply of two bullion stitches. The bullion knots at the side of the central stalk are curled by taking up in the first instance only the smallest piece of the stuff. [77]

to work G.

To work French Knots (G), having brought out your needle at the point where the knot is to be, hold the thread under your thumb, and, letting it lie to the right, put your needle under the stretched part of it. Turn the needle so as to twist the thread once round it. That done, put the needle in again about where it came out, draw it through from the back, and bring it out where the next knot is to be.

For large knots use two or more threads of silk, and do not twist them more than once. With a single thread you may twist twice, but the result of twisting three or four times is never happy.

the working of G on knot-stitch sampler.

The use of knots is shown to perfection in Illustration 24. Worked there in white silk floss upon a dark purple ground, they are quite pearly in appearance, whether in rows between the border lines, or scattered over the ground. They are most useful in holding the design together, giving it mass, and go admirably with chain-stitching, to which, when close together, they have at first sight some likeness. A single line of knots may almost be mistaken for chain-stitch; but of themselves they do not make a good outline, lacking firmness. A happier use of them is to fringe an [78] outline, as for example in the peacock's tail on page 38; but this kind of thing must be used with reticence, or it results in a rather rococo effect. Good use is sometimes made of knots to pearl the inner edge of a pattern worked in outline, or to pattern the ornament (instead of the ground) all over. Differencing of this kind may be an afterthought—and a happy one—affording as it does a ready means of qualifying the colour or texture of ground, or pattern, or part of either, which may not have worked out quite to the embroiderer's liking.

The obvious fitness of knots to represent the stamens of flowers is exemplified in Illustration 93. Worked close together, they represent admirably the eyes of composite flowers, as on the sampler; they give, again, valuable variety of texture to the crest of the stork in Illustration 85.

The effect of knotting in the mass is shown in Illustration 31, embroidered entirely in knots, contradicting, it might seem, what was said above about its unfitness for outline work. The lines, even the voided ones, are here as sharp as could be; but then, it is not many of us who work, knot by knot, with the marvellous precision of a Chinaman. His knotted texture is not, however, always what it seems. He has a way of producing a knotted line by first knotting his thread (it may be done with a netting needle), and then stitching it down on to the surface of the material, which gives a [80] pearled or beaded line not readily distinguishable from knot stitch.

larger image

31. A TOUR DE FORCE IN KNOTS.

The Japanese embroiderer, instead of knotting his own thread, employed very often a crinkled braid. This is shown in the cloud work in Illustration 85. The only true knotting there is in the top-knot of the bird.

32. INTERLACING-STITCH SAMPLER.

33. INTERLACING-STITCH SAMPLER (BACK).

INTERLACINGS, SURFACE STITCHES, AND DIAPERS. [83]

The samplers so far discussed bring us, with the exception of Darning, Satin-stitch, and some stitches presently to be mentioned, practically to the end of the stitches, deserving to be so called, generally in use.

By combining two or more stitches endless complications may be made; and there may be occasions when, for one purpose or another, it may be necessary, as well as amusing, to invent them. In this way stitches are also sometimes worked upon stitches, as shown on the sampler, Illustration 32. You will see, on referring to the back of it (33), that only the white silk is worked into the stuff: the dark is surface work only. There is no end to such possible INTERLACINGS. Those on the sampler do not need much explanation; but it may be as well to say that A starts with crewel-stitching; B and C with back-stitching; D with chain-stitching; E with darning or running; F, G, and H with varieties of herring-bone-stitch; J with Oriental-stitch; and K with feather-stitch. The interlacing on the [84] surface of these is shown in darker silk. C and G undergo a second course of interlacing.

The danger of splitting the first stitches in working the interlacing ones, is avoided by passing the needle eye-first through them.

Other surface work, sometimes called LACE-STITCH, is illustrated in the sampler, Illustration 34. There is really no limit to patterns of this kind. Some are better worked in a frame, but that is very much a matter of personal practice.

the working of F on interlacing-stitch sampler.

to work H, 34.

In the Surface Darning at H (34) long threads are first carried from edge to edge of the square, there only piercing the stuff, and then darned across by other stitches, again only piercing it at the edges.

An oblique version of this is given at C (34).

to work B, 34.

The Lace Buttonholing at B (34) is worked as follows: — Buttonhole three stitches into the stuff from left to right, not quite close together, and further on three more; then, working from right to left, make three buttonhole stitches into the thread connecting the stitch groups; but do not stitch into the stuff except at the ends of the rows. The last row must, of course, be worked into the stuff again.

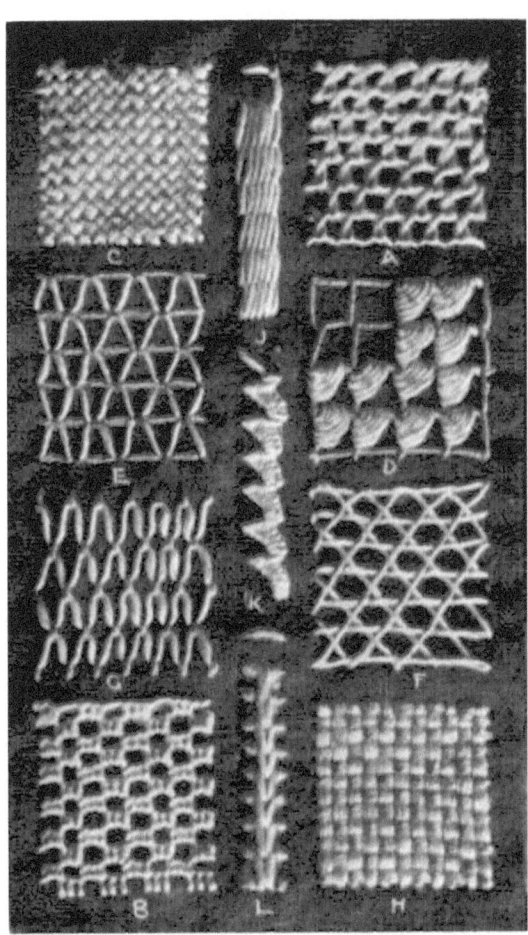

larger image

34. SURFACE-STITCH SAMPLER.

to work F, 34.

[86] Net Passing, as at F (34), is not very differently worked from A or B. It is much more open, and the first row of horizontal stitches is crossed by two opposite rows of oblique stitches, which are made to interlace.

to work G, 34.

The square at G is worked by first making rows of short upright stitches worked into the stuff, and then threading loose stitches through them.

to work D, 34.

The square at D is worked on the open lattice shown; the solid parts are produced by interlacing stitches from side to side, starting at the angle.

In the square at E (Japanese Darning) horizontal lines are first darned, and then zigzag lines are worked between them, much as in G; but, as they penetrate the material, this is scarcely a surface stitch.

to work A, 34.

The horizontal lines at top and bottom of the square at A are back-stitching, the intermediate ones simply long threads carried from one side to the other; they are laced together by lines looped round them.

to work L, 34.

The band at L is begun by making horizontal bar stitches. A row of crewel-stitch and one of outline-stitch, worked on to the bars, and not into the stuff, makes the central chain.

to work K, 34.

The band at K is merely surface buttonholing over a series of slanting stitches.

to work J, 34.

The band at J is buttonhole stitching wide apart, the bars filled in with surface crewel-stitch.

larger image

35. LACE OR SURFACE STITCH.

Most delicate surface stitching occurs in Illustration [88] 35, the fine net being worked only from edge to edge of the spaces it fills, and not elsewhere entering the stuff; which accounts for most of it being worn away. The flower or scroll-work is *bonâ fide* embroidery, worked through the stuff. The delicate network of fine stitching, which once covered the whole of the background, is for the most part neither more nor less than a floating gossamer of lacework.

One cannot deny that that is embroidery, though it has to be said that *lace-stitches* are employed in it.

Stern embroiderers would like to deny it. Of course it is frivolous, and in a sense flimsy, but it is also delicate and dainty to a degree. It is suited only to dress, and that of the most exquisite kind. A French marquise of the Regency might have worn it, and possibly did wear it, with entire propriety — if the word is not out of keeping with the period.

The frailty of this kind of thing is too obvious to need mention, and that, of course, is a strong argument against it.

All attempt to give separate names to diapers of this kind, whether worked upon the surface or into the stuff, is futile. They ought not even to be called stitches, being, in fact, neither more nor less than stitch patterns, to which there is no possible limit, unless it be the limit of human invention. Every ingenious workwoman will find out patterns of her own more or less. They are very useful for [89] filling in surfaces (pattern or background) which it may be inexpedient to work more solidly.

The greater part of such patterns are geometric (Illustrations 35 and 73), following, that is to say, the mesh of the material, and making no secret of it. On Illustration 3 you see very plainly how the rectangular diaperings are built up geometrically on the square lines of the mesh, as was practically inevitable working on such a ground. The relation of stitch to stuff is here obvious.

The choice of stitch patterns of this kind is invariably left to the needlewoman. The utmost a designer need do is to indicate on his drawing that a "full," "open," or "intermediate" diaper is to be used. And the alternation of lighter and heavier diapers should be planned, and not left altogether to impulse, though the pattern may be. Moreover, there is room for the exercise of considerable taste in the choice of simpler or more elaborate patterns, freer or more geometric. Many a time the shape of the space to be filled, as well as its extent, will suggest the appropriate ornament. The diaper design is not, of course, drawn on the stuff, but points of guidance may be indicated through a kind of fine stencil plate.

The patterns used for background diapering need not, as a rule, be intrinsically so interesting as those which diaper the design itself, nor are they usually so full. They take more often the form of spot or sprig patterns, not continuous, in which the [90] geometric construction is not so obvious, nor even necessary. In either case the prime object of the stitching is not so much to make ornamental patterns as to give a tint to the stuff without entirely hiding it with work; and the worker chooses a lighter or heavier diaper according to the tint required. If the work is all in white it is texture, instead of tint, that is aimed at.

For a background, simple darning more or less open, in stitches not too regular, is often the best solution of the difficulty. The effect of the ground grinning through is delightful.

SATIN-STITCH AND ITS OFFSHOOTS. [91]

Satin-stitch is *par excellence* the stitch for fine silkwork. I do not know if the name of "satin-stitch" comes from its being so largely employed upon satin, or from the effect of the work itself, which would certainly justify the title, so smooth and satin-like is its surface. Given a material of which the texture is quite smooth and even, showing no mesh, satin-stitch seems the most natural and obvious way of working upon it. In it the embroidress works with short, straight strokes of the needle, just as a pen draughtsman lays side by side the strokes of his pen; but, as she cannot, of course, leave off her stroke as the penman does, she has perforce to bring back the thread on the under side of the stuff, so that, if very carefully done, the work is the same on both sides.

Satin-stitch, however, need not be, and never was, confined to work upon silk or satin. In fact, it was not only worked upon fine linen, but often followed the lines of its mesh, stepping, as in Illustration 9, to the tune of the stuff. This may be described as satin-stitch in the making—at any [92] rate, it is the elementary form of it, its relation to canvas-stitch being apparent on the face of it. Still, beautiful and most accomplished work has been done in it alike by Mediæval, Renaissance, and Oriental needleworkers.

to work A, 36.

To cover a space with regular vertical satin stitches (A on the sampler, Illustration 36), the best way of proceeding is to begin in the centre of the space and work from left to right. That half done, begin again in the centre and work from right to left.

In order to make sure of a crisp and even edge to your forms, always let the needle enter the stuff there, as it is not easy to find the point you want from the back.

In working a second row of stitches, proceed as before, only planting your needle between the stitches already done. Fasten off with a few tiny surface stitches and cut off the silk on the right side of the stuff: it will be worked over.

to work B, 36.

To cover a space with horizontal satin stitches (B on sampler), begin at the top, and work from left to right. The longer stretches there are not, of course, crossed at one stitch; they take several stitches, dovetailed, as it were, so as not to give lines.

The easiest, most satisfactory, and generally most effective way of working flat satin stitch is in oblique or radiating lines (C, D, E), working in those instances, as in the case of A, from the [95] centre, first from left to right and then from right to left.

larger image

36. SATIN-STITCH SAMPLER.

larger image

37. SATIN-STITCH SAMPLER (BACK).

Stems, narrow leaflets, and the like, are best worked always in stitches which run diagonally and not straight across the form.

In the case of stems or other lines curved and worked obliquely, the stitches must be very much closer on the inner side of the curve than on the outside: occasionally a half-stitch may be necessary to keep the direction of the lines right, in which case the inside end of the half-stitch must be quite covered by the stitch next following.

larger image

38. SATIN-STITCH IN COARSE TWISTED SILK.

Satin-stitch is seen at its best when worked in floss. Coarse or twisted silk looks coarse in this stitch, as may be seen by comparing the petal D in the sampler, Illustration 36, with the petal in twisted silk here given (38). Marvellously skilful as are the needle-workers of India (Illustration 39), they get rather broken lines when they work in thick twisted silk. The precision of line a skilled worker can get in floss is wonderful. An Oriental will get sweeping lines as clean and firm as if [96] they had been drawn with a pen, and this not merely in the case of an outline, but in voided lines of which each side has to be drawn with the needle. The voided outline, by the way, as on Illustrations 39, 40, is not only the frankest way of defining form, but seems peculiarly proper to satin-stitch; and it is a test of skill in workmanship: it is so easy to disguise uneven stitching by an outline in some other stitch. The voiding in the wings of the birds in Illustration 40 is perfect; and the softening of the voided line, at the start of the wing in one case and the tail in the other, by cross stitching in threads comparatively wide apart, is quite the right thing to do. It would have been more in keeping to void the veins of the lotus leaves than to plant them on in cord.

Satin-stitch must not be too long, and it is often a serious consideration with the designer how to break up the surfaces to be covered so that only shortish stitches need be used. You might follow the veining of a leaf, for example, and work from vein to vein. But all leaves are not naturally veined in the most accommodating manner. Treatment is accordingly necessary, and so we arrive at a convention appropriate to embroidery of this kind. It takes a draughtsman properly to express form by stitch distribution. The Chinese convention in the lotus flowers (Illustration 40) is admirable.

larger image

39. SATIN-STITCH IN FINE TWISTED SILK.

It is the rule of the game to lay satin-stitch very evenly. Worked in floss, the mere surface of [98] satin-stitch is beautiful. A further charm lies in the way it lends itself to gradation of colour. Beautiful results may be obtained by the use of perfectly flat tints of colour, as in Illustration 40; but the subtlest as well as the most deliberate gradation of tint may be most perfectly rendered in satin-stitch.

TO WORK SURFACE SATIN-STITCH.

Surface Satin-stitch (not the same on both sides), though it looks very much like ordinary satin-stitch, is worked in another way. The needle, that is to say, after each stitch is brought *immediately* up again, and the silk is carried back on the upper instead of the under side of the stuff. Considerable economy of silk is effected by thus keeping the thread as much as possible on the surface, but the effect is apt to be proportionately poorer. Moreover, the work is not so lasting as when it is solid. The satin-stitch on Illustration 58 is all surface work. It looks loose, which it is always apt to do, unless it is kept stretched on the frame, on which, of course, satin-stitch is for the most part worked. Very effective Indian work is done of this kind—loose and flimsy, but serving a distinct artistic purpose. It is to embroidery of more serious kind what scene painting is to mural decoration.

larger image

40. CHINESE SATIN-STITCH.

Embroidery is often described as being in "long-and-short-stitch," a term properly descriptive not of a stitch, but of its dimensions. Whether you use stitches of equal or of unequal length is a [100] question merely of the adaptation of the stitch to its use in any given instance; there is nothing gained by calling an arrangement of alternating stitches, "long and short," or by calling them "plumage-stitch," or, which is more misleading, "feather-stitch," when they radiate so as to follow the form, say, of a bird's breast. The bodies of the birds in Illustrations 40 and 85 are in plumage-stitch so called. This adaptation of stitch to bird or other forms gives the effect of fine feathering perfectly. But why apply the term "satin-stitch" exclusively to parallel lines of stitches all of a length?

"Long-and-short-stitch," then, is a sort of satin-stitch; only, instead of the stitches being all of equal length, they are worked one *into* the others or *between* them, as in the faces in Illustrations 79 and 80.

A little further removed from satin-stitch is what is known as "split-stitch," in which the needle is brought up *through* the foregoing stitch, and splits it. The way of working this stitch is more fully given on page 105.

The worker adapts, as a matter of course, the length of the stitch to the work to be done, directing it also according to the form to be expressed, and so arrives, almost before he is aware of it, by way of satin-stitch, at what is called plumage-stitch.

larger image

41. OFFSHOOTS FROM SATIN AND CREWEL STITCHES.

larger image

42. OFFSHOOTS FROM SATIN AND CREWEL STITCHES (BACK).

The distinction between the stitches so far [103] described is plain enough, and an all-round embroidress learns to work them; but workers end in working their own way, modifying the stitch according to the work it is put to do, and produce results which it would be difficult to describe and pedantic to find fault with. Even short, however, of such individual treatment, the mere adaptation of the stitch to the lines of the design removes it from the normal. It makes a difference, too, whether it is worked in a frame or in the hand: in the one case you see more likeness to one stitch, in the other to another. The flower at B, for example, and the leaf at D, on the sampler, Illustration 41, are both worked in what is commonly called "plumage," or "embroidery" stitch, though the term "dovetail," sometimes used, seems to describe it better. Instance B, however, is worked in the hand, and D in a frame—from which very fact it follows that the worker is naturally disposed to regard B as akin to crewel-stitch and D to satin-stitch, between [104] which two stitches "dovetail" may be regarded as the connecting link.

the working of B on sampler 41.

to work B, 41.

The petals at B are worked in the method illustrated in the diagram overleaf. The first step is to edge the shape with satin-stitches in threes, successively long, shorter, and quite short. This done, starting at the base again, you put your needle in on the upper or right side of the first short stitch, and bring it out through the long stitch (as shown in the diagram). You then make a short stitch by putting your needle downwards through the material, and taking up a small piece of it. You have finally only to draw the needle through, and it is in position to make another long stitch. As the concentric rings of stitching become smaller, you make, of course, shorter stitches, and you need no longer pierce the thread of the long stitch.

to work D, 41.

The working of the scroll at D on the sampler, Illustration 41, needs no detailed explanation. Anyone who is acquainted with the way satin-stitch is worked (it has already been sufficiently explained), and has read the above account of the working of B, will understand at once how that is worked in the frame.

It will be seen that there is a slight difference in effect between the two, arising from the fact that work done in the hand is necessarily more loosely and not quite so evenly done as that on a frame. [105]

to work split-stitch C, 41.

Split-stitch (C on the sampler), again, resembles either crewel-stitch or satin-stitch, according as it is worked in the hand or on a frame. In working in the hand, you take a rather shorter stitch back than in crewel-stitch, piercing with the needle the thread which is to form the next stitch. In working on a frame, you bring your needle always up through the last-made satin-stitch in order to start the next. Whichever way it is done, split-stitch is often difficult to distinguish without minute examination from chain-stitch. Further reference to its use is made in the chapter on shading. It may be interesting to compare it with crewel-stitch (A on the sampler), which is also a favourite stitch for shading.

DARNING. [106]

It is the peculiarity of Darning and Running that you make several stitches at one passing of the needle.

Darning and running amount practically to the same thing. Darning might be described as consecutive lines of running. The difference is, in the main, a matter of multiplication; but the distinction is sometimes made that in running the stitches may be the same length on the face as on the reverse of the stuff, whereas in darning the thread is mainly on the surface, only dipping for the space of a single thread or so below it.

It results from the way of working that you get in darning an interrupted line characteristic of the stitch. What is called "double darning," by which the breaks in the single darning are made good, has in effect no character of darning whatever.

Darning has a homely sound, but it is useful for more than mending. In embroidery you no longer use it to replace threads worn away, but build up upon the scaffolding of a merely serviceable material what may be a gorgeous design in silk.

larger image

43. DARNING SAMPLER.

[108] Darning is worked, of course, in rows backwards and forwards; but if the stitches are long and in the direction of the weft, it is as well not to run the returning row next to the one just done, but to leave space for a second course of darning afterwards between the open rows.

The darning of the sampler, Illustration 43, is very simple. The flower is darned in stitches of fairly equal length, taking up one thread of the material, and covering a space of almost a quarter of an inch before taking up the next thread. The outline of a petal is first worked, and successive rows of darning follow the lines of the flower, expressing to some extent its form. Much depends upon the direction of the stitch.

The texture of the work depends upon the length of the stitches, and on the amount of the stuff showing through.

Darning is usually supplemented by outlining. The sampler is designed to show how far one can dispense with it. The flower stalk is defined by darning the first row in a darker colour; for the rest, voiding is employed, but it is not easy to void in darning.

The background is darned diaper fashion. It gives, that is to say, deliberately diagonal lines. A background irregularly darned should be irregular enough never to run into lines not contemplated by the worker.

44. DARNING DESIGNED BY WILLIAM MORRIS.

In the case of large leaves, veined, the veining [110] should be worked first, the stitches between them radiating outwards to the edge of the leaf.

More accomplished work in darning is shown in the border by William Morris in Illustration 44, where it appears, however, much flatter than in the coloured silk. It is worked solid, the radiating stitches accommodating themselves to the forms of the leaves and petals, which, in fact, are designed with a view to their execution in this way. They are defined by outline-stitching—light or dark as occasion seemed to require.

Mention has already been made of darning *à propos* of canvas-stitch; and there is a sort of natural correspondence between the *mécanique* of darning in its simplest form and the network of open threads which gives to rectangular darning, like the German work in Illustration 45, character which more than compensates for its angularity in outline. The darning is there quite even in workmanship, but it is, as will be seen, of different degrees of strength—lighter for the surface of the pattern, heavier for the outline.

You may qualify the colour of a stuff by lightly darning it with silk of another shade, and very subtle tints may be got by thus, as it were, veiling a coloured ground with silks of various hues.

larger image

45. FLAT DARNING UPON A SQUARE MESH.

LAID-WORK. [112]

The necessity for something like what is called "Laid-work" is best shown by reference to satin-stitch. It was said in reference to it that satin-stitches should not be too long. There is a great deal of Eastern work in which surface satin-stitch, or its equivalent, floats so loosely upon the face of the stuff that it can only be described as flimsy. Nothing could be more beautiful in its way than certain Soudanese embroidery, in which coloured floss in stitches an inch or more long lies glistening on the stuff without any interruption of threads to fasten it down.

Embroidery of this kind, however, hardly comes within the scope of practical work. Long, loose stitches want sewing down. Some compromise has to be made between art and beauty. The problem is to make the work strong enough without seriously disturbing its lustrous surface, and the solution of it is "laid-work," at which we arrive thus almost by necessity.

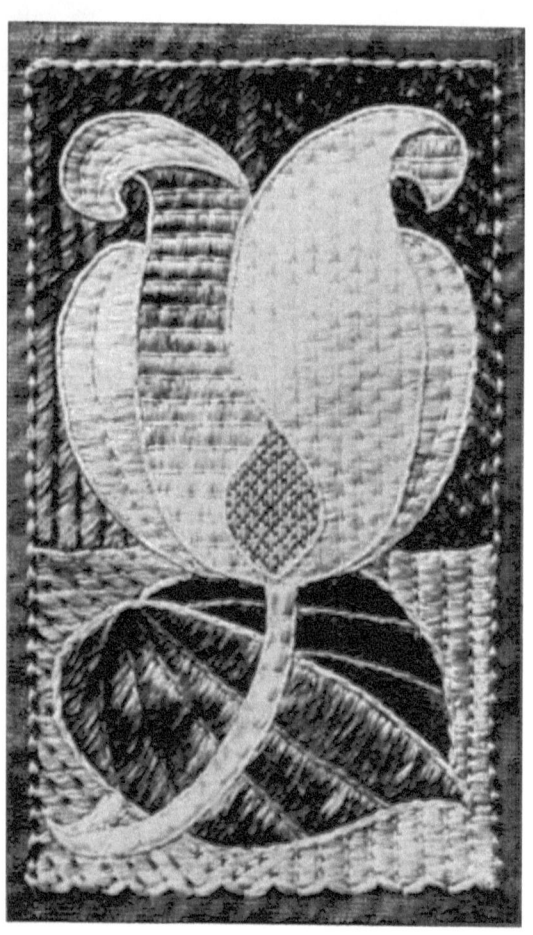

larger image

46. LAID-WORK SAMPLER.

It involves no new stitch, but is only another way of using stitches already described. In laid-work, long tresses of silk, as William Morris called [114] them, floss by preference, are thrown backwards and forwards across the face of the stuff, only just piercing it at the edges of the forms, and back again. These silken tresses are then caught down and kept, I will not say close to the ground, but in their place upon it, by lines of stitching in the cross direction.

Laid-work is not, at the best, a very strong or lasting kind of embroidery (it needs to be carefully covered up even as it is worked), but by no other means is the silky beauty of coloured floss so perfectly set forth. It is hardly worth doing in anything but floss.

Laid-work lends itself also to gradation of colour within certain limits—the limits, that is to say, of the straight parallel lines in which the silk is laid: the direction of these is determined often by the lines of sewing which are to cross them. In any case the direction of the threads is here more than ever important. The sewing down must take lines and may form patterns.

The sampler, Illustration 46, wants little or no explanation. It illustrates the various ways of laying. In the leaf the floss is sewn down with split-stitch, which forms the veining. Elsewhere it is kept in place by "couching," a process presently to be described. For the outlines, split-stitch and couching are employed. The last row of laid work in the grounding is purposely pulled [116] out of the straight by the couching in order to give a waved edge. The diaper which represents the seeding of the flower is not, properly speaking, laid-work: single threads of white purse silk are there couched down with dark.

larger image

47. JAPANESE LAID-WORK.

For the transverse stitching, for which also it is best to use floss, either split-stitch may be used, as in the leaf in the sampler, Illustration 46, or a thread may be laid across and sewn down—couched, as it is called—as in the flower. The closer the cross lines the stronger the work, but the less lustrous the effect.

Laid floss may be employed to glorify the entire surface of a linen material, as in the sampler or for the pattern only upon a ground worth showing, as in Illustrations 47, 48, 49.

Laid-work will not give anything like modelling, and it is not best suited to figure design except where it is quite flatly treated. An instance of its use in figure work occurs on Illustration 79. It is effective when quite naively and simply used in cross lines which do not appear to take any account of the forms crossed—as, for example, in Illustration 47, where the stitching does not pretend to express more than a flat surface. The floss, however, is there carefully laid at a different angle of inclination in each petal, so as to give variety of colour. The lines of sewing vary according to the lines of the laid floss, but do not cross them at right angles. The important thing is, of course, that they should [118] catch the laid "tresses" at intervals not too far apart. If the lines which sew down the floss have also to express drawing, as in the case of the bird's wings in Illustration 48, the underlying floss must be laid in lines which they will cross. In the case of the leaves in the same piece of work, the floss is laid in the direction in which the leaf grows, and the stitching across, which sews it down, is slightly curved so as to suggest roundness in them.

larger image

48. INDO-PORTUGUESE LAID-WORK.

A more finished piece of work is shown in Illustration 49, where the laid floss crosses the forms, and the sewing down takes very much the place of veining in the flower, and of ribs in the scroll, expressing about as much modelling as can be expressed this way, and more, perhaps, than it is advisable often to attempt.

The sewing down asserts itself most, of course, when it is in a colour contrasting with the laid floss, as it does in the leaves in the smaller sampler overleaf.

The stitching down makes usually a pattern more or less conspicuous. On this same sampler it does so very deliberately in the case of the broad stalk. The rather sudden variation of the colour shown there in the leaves is harmless enough in bold work, to which the process is best suited. One may be too careful in gradating the tints: timidity in this respect prevails too much among modern needlewomen: an artist in floss should not want her work to look like a gradated [120] wash of colour. The Italians of the 16th and 17th centuries (see Illustration 49) were not afraid of rather abrupt transition in the shades of colour they used for laid-work.

49. ITALIAN LAID-WORK.

50. LAID SAMPLER.

When laid floss is kept in place by threads themselves sewn down across it, such threads are called "couched," and the work itself may be described as laid and couched. Hence arises some confusion between the two methods of work—laying and couching. It saves confusion to make a sharp distinction between the two—using the term "laid" only for stitches (floss) first loosely laid upon the surface of the stuff and then sewn down by cross lines of stitching of [121] whatever kind, and "couched" for the sewing down of cords, &c. (silk or gold), thread by thread or in pairs. Laid floss is sewn down *en masse*, couched silk in single or double threads; and accordingly laid answers best for surface covering, couched for outlining, except in the case of gold, which even for surface covering is always couched.

COUCHING [122]

Couching is the sewing down of one thread by another — as in the outline of the flower on the laid sampler, Illustration 46. The stitches with which it is sewn down, thread by thread, or, in the case of gold, two threads at a time, are best worked from right to left; or, in outlining, from outside the forms inwards, and a waxed thread is often used for the purpose. Naturally the cord to be sewn down should be held fairly tightly in place to keep the line even.

It is usual in couching to sew down the silk or cord with stitches crossing it at right angles, except in the case of a twisted cord, which should be sewn down with stitches in the direction of the twist.

Couching is best done in a frame; but it may be done in the hand by means of buttonhole-stitch.

larger image

51. A. bullion. B. couched cord.

When a surface is covered with couching, as in the seeding of the flower in the sampler, Illustration 46, the sewing down stitches make a pattern—all the plainer there, because the stitching is in a contrasting shade of colour. It is quite [124] permissible to call attention to the stitching if it suits your artistic purpose. To disguise it by sewing *through* the cord is not a workmanlike practice. A worker should frankly accept a method of work and get character out of it.

Embroidresses have a clever way of untwisting a cord before each stitch and twisting it again after stitching through it—between the strands, that is to say, in which the stitching is lost. The device is rather too clever. It shows a cord with no visible means of attachment to the ground, which is not desirable, however much desired. There is no advantage in attaching cords to the surface of silk so that they look as if they had been glued on to it. Conjuring tricks are highly amusing, but one does not think very highly of conjurers. Personally, I would much rather have seen more plainly the way the cord is sewn down in the graceful cross in Illustration 51, a design perfectly adapted to couching, and yet unlike the usual thing.

Where it is softish silk which is stitched down, it makes a great difference whether it is loosely held and tightly sewn, or the contrary. Contrast the short puffy lines nearest the corners in the sampler, Illustration 52, with the longer ones between the broad and narrow bands. The broad band is worked in rows of double filoselle, of various shades, sewn down with single filoselle. In the narrower bands twisted silk is sewn down with [125] stitches in the direction of its twist. This is more plainly seen in the upper of the two bands, where the sloping stitches are lighter in colour than the cord sewn down.

larger image

52. COUCHING SAMPLER.

Characteristic use is made of rather puffy couching in the ornament of the lady's dress in Miss Keighley's panel, Illustration 61, where it has very much the richness of embroidery in seed pearls.

It was a common practice in Germany in the 16th century to work in solid couching upon cloth, employing a twisted thread and sewing it with stitches in the direction of the twist, so that at first sight one does not recognise it as couching. It looks like rather coarse stitching in the direction [126] of the forms, and expresses shading very well. The cloth ground accounts, perhaps, for the choice of method: the material is not otherwise a pleasant one to embroider upon.

A rather earlier German method was to couch in parallel lines of white upon white linen, and so get relief and texture but no modelling, though the drawing was helped by varying the direction of the parallel lines.

The entire surface of a linen ground was sometimes covered with couched threads of silk or fine wool—some of it in vertical and horizontal lines, some of it in the direction of the pattern. This, again, was a German practice, as may be seen in the Hildesheim Cope at South Kensington.

All-over couching may be used with advantage to renew the ground of embroidery so worn as to be unsightly; and is more lasting than laid-work for the purpose. It is laborious to do, but more satisfactory when done than remounting; and one or the other is a necessity sometimes. The effect of age is, up to a certain point, pleasing: rags are not.

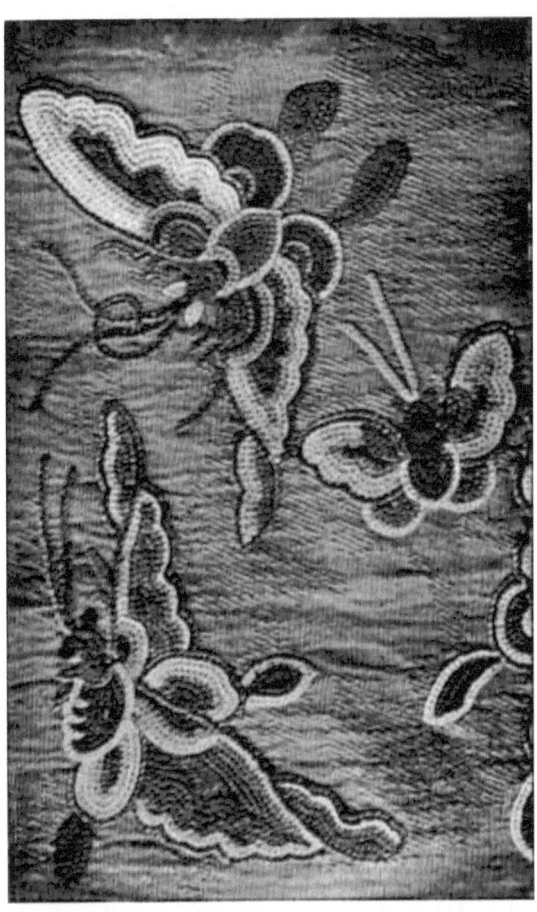

larger image

53. COUCHING IN LOOPED THREADS.

Couching, however (except with gold), was more commonly used for outlining, and is quite peculiarly suited to give a firm line. A beautiful example of outline work in coloured silk upon white linen is pictured in Illustration 90, in which the lines of delicate Renaissance arabesque are perfectly preserved. The rare practice of such work as [128] this, notwithstanding its distinction, is perhaps sufficiently accounted for by its modesty. It is true, it wants well-considered and definitely drawn design, and there is no possible fudging with it.

larger image

54. REVERSE COUCHING.

The value of a couched cord as an outline to stitching (satin-stitch in this instance) is shown in Illustration 91, in which the singularly well-schemed and well-drawn lines of the ornament are given with faultless precision. This is a portion of an altogether admirable frame to an altogether foolish picture in needlework, of which a fragment only is shown.

The appropriateness of couched cord to the outlining of inlay or of appliqué is seen in the two examples which form Illustration 62. In the one (A) it defines the clear-cut counterchange pattern; in the other (B), being of a tint intermediate between the ground and the ornament, it softens the contrast between them. An interesting technical point in the design of this last is the way the cord outlining the leaves makes a sufficiently thick stalk, coming [129] together, as it naturally does, double at the ends of the leaves.

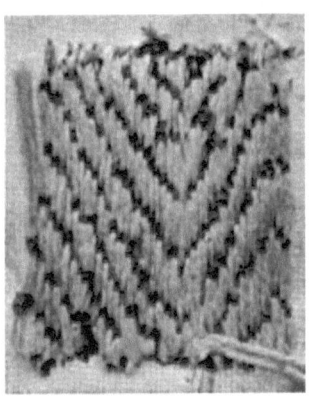

larger image

55. REVERSE COUCHING (BACK).

This occurs again in Illustration 63, where the double threads which form the stalks, though separately stitched down, are couched again at intervals by bands crossing the two—at the springing of the stalks and tendrils, for example, where joins inevitably occur. The cords forming the central stalk are in one case looped.

Fantastic use has often been made of the looping of couched cord. The Spanish embroiderers made most ornamental use of a wee loop at the points of the leaves where the cord must turn; but the device of looping may easily be used to frivolous purpose. A regularly looped line at once suggests lace. A perplexing Chinese practice is to couch fine cord in little loops so close together that they touch. A surface filled in after this manner, as in the butterflies on Illustration 53, might pass at first sight for French knots or chain-stitch: it is really another method of all-over couching.

A double course of couching forms the outline in [130] Illustration 92, one of filoselle and one of cord, separately sewn; but the tendrils, which are of silver thread, are sewn down both threads at a time with double stitches, very obvious in the illustration. Over the couched silver threads which form the main rib of the leaf a pattern is stitched in silk.

A propos of couching, mention must be made of a way of working used in the famous Syon Cope by way of background, and figured overleaf (Illustration 54). The ground stuff is linen, twofold, and it is worked in silk, which lies nearly all upon the surface. The stitch runs from point to point of the zigzag pattern; there it penetrates the stuff, is carried round a thread of flax laid at the back of the material, and is brought to the surface again through the hole made by the needle in passing down. That is to say, the silken thread only *dips* through the linen at the points in the pattern, and is there caught down by a thread of flax on the under-surface of the linen. The reverse of the work (Illustration 55) shows a surface of flax threads couched with silk, for which reason the method may be described as reverse couching. On the face it gives an admirable surface diaper, flat without being mechanical. It is easily worked with a blunt needle; with a sharp one there would be a danger of splitting the stitch. It is a kind of work on which two persons might be employed, one on either side of the stuff.

COUCHED GOLD. [131]

In olden days silk does not appear to have been couched in the East. On the other hand, it was the custom to couch gold thread in Europe at least as early as the twelfth century; so that the method was probably first used for gold, which, except in the form of thin wire or extraordinarily fine thread, is not quite the thing to stitch with. Besides, it was natural to wish to keep the precious metal on the surface, and not waste it at the back of the stuff.

A distinguishing feature about gold is that by common consent it is used double and sewn down two threads at a time. This is not merely an economy of work; but, except in the case of thick cords or strips of gold, it has a more satisfactory effect—why it is not easy to say. Panels A, B, C, in the sampler, Illustration 56, are couched in double threads, D in single cords.

Gold couching is there used, as it mostly is, to cover a surface. In doing that, it is usual to sew the threads firmly down at the edges of the forms and cut them very sharply off; but they may equally well be carried backwards and forwards [132] across the face of the stuff. The slight swelling of the gold thread where it turns gives emphasis to the outline; but the turning wants carefully doing, and the gold thread must not be too thick. If you use a large needle (to clear the way for the thread), the turning of the gold may take place on the back instead of on the face of the material, but only in the case of very fine thread.

Gold threads often want stroking into position. This may be done with what is called a "pierce"; but a good stiletto, or even a very large needle, will answer the purpose. Sharply pointed scissors are indispensable.

In solid couching the stitches run almost inevitably into pattern; and it is customary, therefore, to start with the assumption that they will, and deliberately to make them into pattern—to work them, that is to say, in vertical, diagonal, or cross lines as at A, in zigzags as at B, or in some more complicated diaper pattern as at C, where the stitching is purposely in pronounced colour, that the pattern may be quite clearly seen; at D it has more its proper value, that the effect of it may be better appreciated. The pattern may, of course, be helped by the colour of the stitching, and there is some art in making the necessary stitches into appropriate pattern.

larger image

56. COUCHED GOLD SAMPLER.

In fact the ornamentist, being an ornamentist, naturally takes advantage of the necessity of stitching, to pattern his metallic surfaces with diaper, [134] using often, as in the scroll in Illustration 57, a diversity of patterns, which gives at once varied texture and fanciful interest to the surface. There is quite an epitome of little diapers in that fragment of needlework; and one can hardly doubt that the embroiderer found it great fun to contrive them. The flat strips of metal emphasising the backs of the curves are sometimes twisted as they are sewn.

The other diapers on the sampler, F, G, H, J, 56, are emphasised by the relief given to them by underlying cords, purposely left bare in parts to show the structure. These underlying cords must be firmly sewn on to the linen ground, and if the stitching follows the direction of the twist in them, the round surface is not so likely to be roughened by it. By rights, the cords should be laid farther apart than in the sampler, where the attempt to force the effect (for purposes of explanation) has not proved very successful. An infinity of basket patterns, as these may be called (basket *stitches* they are not), may be devised by varying the intervals at which the gold threads are sewn down, and the number of cords they cross at a time.

larger image

57. COUCHED SILVER.

The central panel of the sampler (E) shows a combination of flat and raised gold. The outline of the heart is corded; the centre of it is raised by stitching, first with crewel wool and then with gold-coloured floss across that (it is difficult to prevent *white* stuffing from showing through [136] gold). This gives only a hint of what may be done in the way of raised ornament upon a flat gold ground, and was done in mediæval work. A single cord may be sewn down to make a pattern in relief, leafage, scrollwork, or what not, which, when the surface is all worked over with gold, has very much the effect of gilt gesso. If, for any reason, heavy work of this kind is to be done on silk or satin, that must first be backed with strong linen.

In mediæval and church work generally the double threads are usually laid close together, forming, as in the diapers on sampler, a solid surface of gold; and that was largely done in Oriental embroidery too—in Chinese, for example, where, however, the threads, instead of being couched in straight lines, follow the outlines of the design, and are worked ring within ring until the space is filled, as in the dragon's face, A, Illustration 58. There is here, as in the working of his body, a certain economy of gold; a small amount of the ground is allowed to show between the lines of double gold thread—not enough to tell as ground, but enough to give a tint of the ground colour to the metal. Further, in this more open couching the direction of the lines of couching goes for more than in solid work. The pattern made by the gold thread is here not only ornamental but suggestive of the scaly body of the creature. It will be seen, too, how, in the working [138] of the legs, the relatively compact gold threads are kept well within the outline, by which means anything like harshness of silhouette is avoided.

larger image

58. COUCHED GOLD NOT QUITE SOLID.

That this less solid manner was not confined to the far East is shown by the Venetian valance, B, on the lower part of the page, which has very much the appearance of gold lace.

A good example of outline (single thread) in gold is given in Illustration 59, part of an Italian housing, which reminds one both in effect and in design of damascening, to which it is in some respects equivalent; only, instead of gold and silver wire beaten into black iron or steel, we have gold and silver thread sewn on to dark velvet. The design recalls also the French bookbindings of the period of

Henri II., in which the tooled ornament was precisely of this character. The resemblance is none the less that an occasional detail is worked more solidly; but, in the main, this is outline work, and a beautiful example of it. The art in work of that kind is, of course, largely in the design. Gold thread work in spiral forms has very much the effect of filagree in gold wire.

The next step is where the cords of gold enclose little touches of embroidery in coloured floss, as in Illustration 91. These have the value of so many jewels or bits of bright enamel. In fact, just as outline work in simple gold thread resembles damascening or filagree, so this outlining of little spaces of coloured silk suggests enamel. [140] The cord of the embroiderer answers to the cloisons of the enameller, the surfaces of shining floss to the films of vitreous enamel.

larger image

59. COUCHED OUTLINE WORK.

Appliqué embroidery is constantly edged with gold or silver thread. An effective, if rather rude, example of this, the thread here again double, is given in Illustration 60.

In couching more than one thread at a time there is a difficulty in turning the angles. The threads give, of necessity, only gently rounded forms. To get anything like a sharp point, you must stop short with the inner thread before reaching the extreme turning point, and take it up again on your way back. What applies to two threads, applies of course still more forcibly to three.

The colour with which gold thread is sewn is a question of considerable importance. If the stitches are close enough together to make solid work, they give a flush of colour to the gold. Advantage is commonly taken of this both in mediæval and Oriental work to warm the tint by sewing it down with red. The Chinese will even work with a deeper and a paler red to get two coppery shades. White stitching pales the gold, yellow modifies it least, green cools it, and blue makes it greener. The closer the stitches, the deeper the tint, of course.

larger image

60. APPLIQUÉ—SATIN ON VELVET.

You can get thus various shades of gold out of the same thread, and even gradation from one to another, as may be seen in a great deal of [142] Spanish work of the 16th century, in which the gold ornament is often quite delicately shaded from yellowish gold to ruddy copper on the one hand, and to bronzy green on the other. Similar use may be made of vari-coloured silks in couching white or other cord; but gold reflects the colour much better than silk, and gives much more subtle effects.

The Flemings and Italians of the early Renaissance went further. They had a way of laying threads of gold and sewing them so closely over with coloured silk that in many parts it quite hid the gold. Only in proportion as they wanted to lighten the colour of the draperies in their pictorial embroideries did they space the stitches farther and farther apart, and let the gold gleam through. Except in the high lights it did not pronounce itself positively. The effect is not unlike what is seen in paintings of the primitive school, where the high lights of the red and blue draperies are hatched with gold. The practice of the embroiderer may be reminiscent of that, or that may be the origin of the primitive painters' convention. It is more as if the embroiderer wanted to represent a precious tissue, a stuff shot with gold.

Illustration 80 gives part of a figure worked in this way, relieved against a more golden architectural background rendered by the very same double threads of gold which run through the figures. In [143] the architecture, however, they are couched in stitches which are never so near as to take away from the effect of the gold. The two degrees of obscuring or clouding gold by oversewing are here shown in most instructive contrast. The cords, as usual, are laid in horizontal courses. That was the convenient way of working; but it resulted in a corded look, which has very much the appearance of tapestry; and there is no doubt that resemblance to tapestry was in the end consciously sought. That the method here employed was laborious needs no saying; but it gave most beautiful, if pictorial, results.

APPLIQUÉ. [144]

Embroidery, it has been shown, is much of it on the surface of the stuff, not just needle stitches, but the stitching-on of something — cord, gold thread, or whatever it may be. And instances have been given where the design of such work was not merely in outline, but where certain details were filled in with stitching. Yet another practice, and one more strictly in keeping with the onlaying of cord, was to onlay the solid also, applying, that is to say, the surface colour also in the form of pieces of silk cut to shape.

Patterns of this kind may be conceived as line work developing into leafy terminations, the Appliqué only an adjunct to couching (Illustration 63); or they may be thought of as massive work eked out with line: the appliqué, that is to say, the main thing, the couching only supplementary (Illustration 92). An intermediate kind is where outline and mass — couching and appliqué — play parts of equal importance in the scheme of design (Illustration 60).

Couched cord or filoselle is useful in covering [145] the raw edge of the onlay, not so much masking the joints as making them sightly.

Appliqué must be carefully and exactly done, and is best worked in a frame. It is almost as much a man's work as a woman's. Embroidery proper is properly woman's work; but here, as in the case of tailoring, the man comes in. The getting ready for appliqué is not the kind of thing a woman can do best.

The finishing may sometimes be done in the hand, and very bold, coarse work may possibly be worked throughout in the hand, and outlined with buttonhole-stitch (chain-stitch is not so appropriate); but when a couched outline is employed it must be done in a frame, and, indeed, work with any pretensions to finish is invariably begun and finished in the frame.

TO WORK APPLIQUÉ

To work appliqué you want, in fact, two frames — one on which to mount the material to be embroidered, and another on which to mount the material to be applied. The backing in each case should be of smooth holland. This is stretched on to the frame, and then pasted with stiff starch or what not; the silk or velvet is laid on to it and stroked with a soft rag until it adheres, and is left to dry gently. When dry, the outlines of the complete design are traced upon the one, and those of the details to be applied upon the other. (You may paste, of course, silks of two or three colours upon one backing for this.) The [146] stuff to be applied is then loosened from its frame, the details are cleanly cut out with scissors, or, better still, a knife (in either case sharp), and transferred to their place in the design on the other frame. There they are kept in position by short steel pins planted upright into the stuff until you are sure they fit, and then tacked firmly down, with care that the stitches are such as will be quite covered by the final couching, chain stitch, or whatever is to be your outline.

In the case of silk or other delicate material, peculiar care must be taken that the paste is not moist enough to penetrate the stuff; but an experienced worker has no fear of that.

A firm outline is a condition of appliqué, and couched cord fulfils it most perfectly. Much depends upon a tasteful and tactful choice of colour for it. You fatten your pattern by outlining it with a colour which goes with it (Illustration 62, B). You thin it by one which goes into the ground. Very subtle use may be made of a double outline or of a corded line upon couched floss. There is a double outline to the ornament in Illustration 92: the inner one next to the yellow satin appliqué is of gold, the outer one next the crimson velvet ground is of white sewn with pale blue. This gives emphasis to the bold forms of the leafage. The mid-rib there is of silver couching; the minor veinings are stitched in silk, and are rather insignificant.

larger image

61. APPLIQUÉ PANEL BY MISS KEIGHLEY.

[148] The less there is of extra stitching on appliqué the better as a rule. It disturbs the breadth, which is so valuable a characteristic of onlay. In no case is much mixing of methods to be desired; but if appliqué is to be supplemented, it had best be with couching, which is not so much stitching as stitched down, itself another form of applied work.

Appliqué of itself is not, of course, adapted to pictorial work, but that in association with judicious stitching and couching it may be used to admirable decorative purpose in figure design is shown by Miss Mabel Keighley's panel, Illustration 61. What an artist may do depends upon the artist. Miss Keighley's panel indicates the use that may be made of texture in the stuff onlaid.

Appliqué is especially appropriate to bold church work, fulfilling perfectly that condition of legibility so desirable in work necessarily seen oftenest from afar. Broadly designed, it may be as fine in its way as a piece of mediæval stained glass, and it gives to silk and velvet their true worth. The pattern may be readable as far off as you can distinguish colour.

larger image

62. A. counterchange. B. appliqué.

Appliqué work is thought by some to be an inferior kind of embroidery, which it is not. It is not a lower but another kind of needlework, in which more is made of the stuff than of the stitching. In it the craft of the needleworker is not carried to its limit; but, on the other hand, [150] it makes great demands upon design. You cannot begin by just throwing about sprays of natural flowers. It calls peremptorily for treatment—by which test the decorative artist stands or falls. Effective it must be; coarse it may be; vulgar it should not be; trivial it can hardly be; mere prettiness is beyond its scope; but it lends itself to dignity of design and nobility of treatment. Of course, it is not popular.

A usual form of appliqué is in satin upon velvet. Velvet on satin (B, Illustration 62) is comparatively rare; but it may be very beautiful, though there is a danger that it may look like weaving.

Silk upon silk (figured damask) is shown in Illustration 63, designed to be seen from a nearer point of view, and less pronounced in pattern accordingly. The strap work, applied in ribbon, is broken by cross stitches in couples, which take away from the severity of the lines. The grape bunches are onlaid, each in one piece of silk, the forms of the separate grapes expressed by couching. The French knots in the centre of the grapes add greatly to the richness of the surface. The leaves are in one piece. It would have been possible to use two or three, joining them at the veins.

63. APPLIQUÉ – SILK ON SILK DAMASK.

The application of leather to velvet, as in Illustration 94, allows modification in the way of execution, and of design adapted to it. Leather does not fray, and needs, therefore, no sewing over at the edge, but only sewing down, which may be done, [152] as in this case, well within the edge of the material, giving the effect of a double outline. The Chinese do small work in linen, making similar use of the stitching within the outline, but turning the cut edge of the stuff under; it would not do to leave it raw. On a bolder scale, but in precisely the same manner, is embroidered the wonderful tent of François Ier., taken at the battle of Pavia, and now in the Armoury at Madrid—obviously Arab work. Something of the kind was done also in Morocco, which points to leather work as the possible origin of this method.

Another ingenious Chinese notion is to sew down little five-petalled flowers (turned under at the edges) with long stamen stitches radiating from a central eye of knots.

INLAY, MOSAIC, CUT-WORK. [153]

A step beyond the process of onlaying is Inlay, where one material is not laid on to the other, but into it, both being perhaps backed by a common material. The process is, in fact, precisely analogous to that inlay of brass and tortoiseshell which goes by the name of its inventor, Boule. The work is difficult, but thorough. It does not recommend itself to those who want to get effect cheaply. The process is suited only to close-textured stuffs, such as cloth, which do not fray.

TO WORK INLAY.

The materials are not pasted on to linen, as in the case of appliqué. The cloth to be inlaid is placed upon the other, and both are cut through with one action of the knife, so that the parts cannot but fit. The coherent piece of material (the ground, say, of the pattern) is then laid upon a piece of strong linen already in a frame; the vacant spaces in it are filled up by pieces of the other stuff, and all is tacked down in place. That done, the work is taken out of the frame, and the edges sewn together. The backing can then, if necessary, be removed; and in Oriental work it generally was. [154]

Inlay lends itself most invitingly to Counterchange in design, as seen in the stole at A, Illustration 62. Light and dark, ground and pattern, are there identical. You cannot say either is ground; each forms the ground to the other. And from the mere fact of the counterchanging you gather that it is inlaid, and not onlaid.

TO WORK COUNTERCHANGE.

Prior to inlaying in materials which are at all likely to fray, you first back them with paper, thin but tough, firmly pasted; then, having tacked the two together, and pinned them with drawing-pins on to a board, you slip between it and the stuff a sheet of glass, and with a very sharp knife (kept sharp by an oilstone at hand) cut out the pattern. What was cut out of one material has only to be fitted into the other, and sewn together as before, and you have two pieces of inlaid work — what is the ground in one forming the pattern in the other, and *vice versâ*. By this ingenious means there is absolutely no waste of stuff. You get, moreover, almost invariably a broad and dignified effect: the process does not lend itself to triviality. It was used by the Italians, and more especially by the Spaniards of the Renaissance, who borrowed the idea, of course, from the Arabs.

larger image

64. INLAY IN COLOURED CLOTHS.

In India they still inlay in cloth most marvellously, not only counterchanging the pattern, but inlaying the inlays with smaller patternwork, thus combining great simplicity of effect with wonderful minuteness of detail. They mask the joins with [156] chain-stitch, the colour of it artfully chosen with regard to the two colours of the cloth it divides or joins. Further, they often patch together pieces of this kind of inlay.

Inlay itself is a sort of Patchwork. You cut pieces out of your cloth, and patch it with pieces of another colour, covering the joins perhaps, as on Illustration 64, with chain stitch, which gives it some resemblance to cloisonné enamel, the cloisons being of chain-stitch.

Where there is no one ground stuff to be patched, but a number of vari-coloured pieces of stuff are sewn together, they form a veritable Mosaic, reminding one, in coloured stuffs, of what the mediæval glaziers did in coloured glass. Admirable heraldic work was done in Germany by this method; and it is still employed for flag making. The stuffs used should be as nearly as possible of one substance. In patchwork of loosely-textured material each separate piece of stuff may be cut large, turned in at the edge, and oversewn on the wrong side.

larger image

65. CUT-WORK IN LINEN.

The relation of Cut-work to inlay is clear—in fact, the one is the first step towards the other. You have only to stop short of the actual inlaying, and you have cut-work. Fill up the parts cut out in Illustration 65 with coloured stuff, and it would be inlay. The needlewoman has preferred to sew over the raw edges of the stuff, and give us a perfect piece of Fretwork in linen. It is part of [158] the game in cut-work to make the fret coherent, whole in itself. The design should tell its own tale. "Ties" of buttonhole-stitch, or what not, are not necessary, provided the designer knows how to plan a fret pattern. Their introduction brings the work nearer to lace than embroidery. The sewing-over may be in chain-stitch, satin-stitch (as in Illustration 65), or in buttonhole-stitch—which last is strongest.

As, in the case of appliqué, inlay, and mosaic, an embroidered outline is usually necessary to cover the join, so in the case of cut-work sewing-over is necessary to keep the edges from fraying. It may sometimes be advisable to supplement this outlining by further stitching to express veining, or give other minute details—just as the glassworker, when he could not get detail small enough by means of glazing, had recourse to painting to help him out. But there is danger in calling in auxiliaries. It is best to design with a view to the method of work to be employed, and to keep within its limits. To worry the surface of applied, inlaid, or cut stuff with finnikin stitchery, is practically to confess either the inadequacy of the design or the fidgetiness of the worker. It should need, as a rule, no such enrichment.

EMBROIDERY IN RELIEF. [159]

Embroidery being work *upon* a stuff, it is inevitably raised, however imperceptibly, above the surface of it. But there is a charm in the unevenness of surface and texture thus produced; and the aim has consequently often been to make the difference of level between ground-stuff and embroidery more appreciable by UNDERLAY or padding of some kind. The abuse of this kind of thing need not blind us to the advantages it offers.

There are various ways of raising embroidery, the principal of which are illustrated on the sampler overleaf.

to work A (66).

In sprig A the underlay is of closely-woven cloth, darker in colour than would be advisable except for the purpose of showing what it is: it is as well in the ordinary way to choose a cloth more or less of the colour the embroidery is to be. The cloth is cut with sharp scissors carefully to shape, but a little within the outline, and pasted on to the linen. When perfectly dry, it is worked over with thick corded silk couched in the ordinary way.

to work B.

The raised line at B reveals the way the stem in Illustration 86 was worked. Two cords of smooth [160] string (macramé, for example) are twisted and tacked in place. Over this floss is worked in close satin-stitch.

to work C.

In sprig C the underlay is of parchment, lightly stitched in place. The use of a double underlay in parts gives additional relief. The embroidery upon this (in slightly twisted silk) is in satin-stitch.

to work D.

The leaf shapes at D are padded with cotton wool, cut out as nearly as possible to the shape required, and tacked down with fine cotton. They are then worked over with floss in satin-stitch. The stalks are not padded with cotton wool, but first worked with crewel wool, which, being soft and elastic, forms an excellent ground for working over in floss silk.

to work E.

In working a stalk like that at E, you first lay down a double layer of soft, thick cotton, and then work over it with flatter cotton (made expressly for padding) in slanting satin-stitch. Three threads of smooth round silk are then attached to one side of the padding and carried diagonally across to the other side, where they are sewn down with strong thread of the same colour close to the underlay, so that the stitches may not show. They are then brought back to the side from which they started, sewn down, and returned again, and so backwards and forwards to the end. The crossing threads make a sort of pattern, and it is a point of good workmanship that [162] they should cross regularly. Such pattern is more obvious when threads of three different shades of colour are employed. Threads of twisted silk may, of course, be equally well used this way without padding underneath.

66. RAISED WORK SAMPLER.

to work F.

In sprig F the underlay is of cardboard, pasted on to the linen. It is worked over with purse silk, to and fro across the forms, and sewn down at the margin with finer silk. This is a method of work often employed when gold thread is used.

to work G.

In sprig G the underlay or stuffing is of string, sewn down with stitches always in the direction of the twist. It is worked over with floss in satin-stitch.

to work H.

In sprig H the underwork consists of stitching in soft cotton, over which thick silk is embroidered in bullion-stitch. The rule is to work the first stitching in such a direction that the surface work crosses it at right angles. The small leaf is worked over with fine purse silk in satin-stitch, which is used also for the stalk.

In the smaller sampler of laid-work, Illustration 50, the broad stem is twice underlaid with crewel, excellent for this soft sort of padding, on account of its elasticity. The leaves have there only one layer of understitching.

Raised work in white upon white is often used for purposes which make it inevitable that sooner or later the work will be washed. That is a consideration which the embroidress must not leave [164] out of account. In any case, work over stitchery is more durable than over loose padding such as cotton wool.

larger image

67. RAISED WORK SHOWING UNDERLAY.

The 15th century work reproduced in Illustration 67 is in flax thread on linen, and the underlay (laid bare in the topmost flower) is of stiff linen, sewn down, not at the margins as in the case of the parchment on the sampler (Illustration 66), but by a row of stitching up the centre of each petal. The veins of the leaves in Illustration 88 are padded with embroidery cotton and worked over with filo-floss. The leaves themselves are not padded, though the sewing down of the veins upon them, as well as the fact that they are applied on to the velvet ground, gives some appearance of relief.

RAISED GOLD. [165]

Our sampler of raised work is done in silk. Underlaying is more often used to raise work in gold, to which in most respects it is best suited. The methods shown in the sampler would answer almost equally well for gold, except that working in gold one would not at H (66) use bullion-stitch, but bullion, first covering the underlay of stitching with smoothly-laid yellow floss.

Bullion consists of closely coiled wire. It is made by winding fine wire tightly and closely round a core of stouter wire. When this central core of wire is withdrawn, you have a long hollow tube of spirally twisted wire. This the embroidress cuts into short lengths as required, and sews on to the silk—as she would a long bead or bugle. Its use is illustrated at A in Illustration 51, where the stems of triple gold cord are tied down at intervals by clasps of bullion, and the leaves, again, are filled in with the same.

It was the mediæval fashion to encrust the robes of kings and pontiffs with pearls and precious stones mounted in gold: the early Byzantine form of crown was practically a velvet [166] cap, on to which were sewn plaques of gorgeous enamel and mounted stones. When to such work embroidery was added, it was not unnatural that it should vie with the gold setting. As a matter of fact, its design was often only a translation into needlework of the forms proper to the goldsmith.

Yet more openly in rivalry with goldsmiths' work was some of the embroidery of the Renaissance, in which the idea—a most mistaken one, of course—seems to have been to imitate beaten metal. This led inevitably to excessively high relief in gold embroidery. You may see in 17th century church work the height to which relief can be carried, and the depth to which ecclesiastical taste can sink.

The Spaniards were, perhaps, the greatest sinners in this respect, seeking, as they did, richness at all cost; but it must be confessed that, in the 16th century at least, they produced most gorgeous results: there is in the treasury of the cathedral at Toledo an altar frontal in gold, silver, and coral, and a yet more beautiful mantle of the Virgin in silver and pearls upon a gold ground, which make one loth to dogmatise.

larger image

68. RAISED GOLD.

The preciousness of gold and silver, points, in the nature of things, to their use for church vestments and the like; and high relief gives, no doubt, value to the metal; but the consideration of its [168] intrinsic value leads quickly to display. The artistic value of gold is not so much that it looks gorgeous as that it glorifies the colour caught, so to speak, in its meshes.

Admitting that there is reason for relief in gold embroidery — it catches the light as flat gold does not — one feels that the very slightest modelling is usually enough. Reference was made (page 136) to the effect of gilt gesso obtained in raised gold thread: that really is about the degree of relief it is safe to adopt in gold embroidery, the relief that is readily got by laying on gesso with a brush, not carving or modelling it; and the characteristically blunt forms got by that means repeat themselves when you work with the needle.

There is ample relief in the gold embroidery on Illustrations 68 and 86. The first of these shows both flat and raised work: the latter illustrates not only various degrees of relief, but several ways of underlaying. It scarcely needs pointing out that the flatter serrated leaves are worked over parchment or paper, and the puffy parts of the flowers over softer padding. Allusion has already been made (page 159) to the way the stalk is worked over twisted cords, as on the sampler, Illustration 66. The patterns in which the gold is worked do not tell quite so plainly here as on Illustration 68, where the basket pattern is more pronounced. In the stalk there flat gold wire is used, and again in the broken surface towards the top of the plate. [169]

Spangles of gold may be used with admirable effect, at the risk, perhaps, of a rather tinselly look; but that has been often most skilfully avoided both in mediæval work and in Oriental. In India great and very cunning use is made of spangles, by the Parsees in particular, who, by the way, embroider with gold wire.

Gold foil may be cut to any shape and sewn on to embroidery, but spangles take mainly one of two shapes, best distinguished as disc-like and ring-like. The discs are flat, pierced in the centre, and sewn down usually with two or three radiating stitches (A, Illustration 51, and Illustration 67). The rings may be attached by a single thread. They can easily be made to overlap like fish scales, and most elaborately embossed pictures have been worked in this way. There is a vestment in the cathedral at Granada which is a marvel to see; but not the thing to do, surely.

Relief is easily overdone, in figure work so easily that one may say safety is to be found only in the most delicate relief. To make figures look round is to make them look stuffed. That stuffy images are to be found in mediæval church work is only too true. In Gothic art one finds this quaint, perhaps, but it is perilously near the laughable. The point of the ridiculous is plainly overpassed in English work of the 17th century, which degenerates at last into mere doll work—the dolls duly stuffed and dressed in most childish [170] fashion, their drapery, in actual folds, projecting. Some really admirable needlework was wasted upon this kind of thing, which has absolutely no value, except as an object-lesson in the frivolity of the Stuarts and their on-hangers.

QUILTING. [171]

A most legitimate use of padding is in the form of Quilting, where it serves a useful as well as an ornamental purpose. To quilt is to stitch one cloth upon another with something soft between (or without anything between). Our word "counterpane" is derived from "contre-poinct," a corruption of the French word for back-stitch, or "quilting" stitch, as it was called.

If you merely stitch two thicknesses of stuff together in a pattern, such as that on Illustration 69, the stuff between the stitches has a tendency to rise: the two layers of stuff do not lie close except where they are held together by the stitching, and a very pleasantly uneven surface results. This effect is enhanced if between the two stuffs there is a layer of something soft. If, now, you keep down the groundwork of your design by comparatively frequent stitches diapering it, you get a pattern in relief, more or less, according to the substance of your padding.

Another way is to pad the pattern only, as in Illustration 70, where the padding is of soft cord.

larger image

69. QUILTING, DONE IN CHAIN-STITCH FROM THE BACK.

A cunning way of padding is first to stitch the [173] outline of the design, and then from the back to insert the stuffing. You first pierce the stuff with a stiletto, and, having pushed in the cord, cotton, or what not, efface as far as possible the piercing: the stuffing has then not much temptation to escape from its confinement.

The Persians do most elaborate quilting on fine white linen, which they sew with yellow silk; but the pattern is stuffed with cords of blue cotton, the colour of which just grins through the white sufficiently to cool it, and to distinguish it from the creamy white ground made warmer by the yellow stitching.

Quilting is most often done in white upon colour, or in one colour upon white. Yellow silk on white linen (as in the case of Illustration 69) was a favourite combination, and is always a delicate one. But there is no reason why a variety of colours should not be used in a counterpane. When you stitch down the ground with coloured silk you give it, of course, colour as well as flatness.

larger image
70. RAISED QUILTING.

STITCH GROUPS. [175]

There are all sorts of ways in which stitches might be grouped:—according to the order of time in which historically they came into use; according as they are worked through and through the stuff or lie mostly on its surface; according as they are conveniently worked in the hand or necessitate the use of a frame; and in other ways too many to mention. It is not difficult, for example, to imagine a classification according to which the satin-stitch in Illustration 71 would figure as a canvas stitch.

In the Samplers they are grouped according to their construction, that seeming to us the most practical for purposes of description. They might for other purposes more conveniently be classed some other way. At all events, it is helpful to group them. Designer and worker alike will go straighter to the point if once they get clearly into their minds the stitches and their use, and the range of each—what it can do, what it can best do, what it can ill do, what it cannot do at all.

Anyone, having mastered the stitches and grasped their scope, can group them for herself, [176] say, into stitches suited (1) to line work, (2) to all-over work, (3) to shading, and so on.

These she might again subdivide. Of line stitches, for example, some are best suited for straight lines, others for curved; some for broad lines, others for narrow; some for even lines, others for unequal; some for outlining, others for veining.

And, further, of all-over stitches some give a plain surface, others a patterned one; some do best for flat surfaces, others for modelled; some look best in big patches, some answer only for small spaces.

With regard to shading stitches, there are various ways (see the chapter on shading) of giving gradation of colour and of indicating relief or modelling.

Some stitches, of course, are adapted to various uses, as crewel, chain, and satin stitches—naturally the most in use. Workers generally end in adopting certain stitches as their own. That is all right, so long as they do not forget that there are other stitches which might on occasion serve their purpose. Anyway, they should begin by knowing what stitches there are. Until they know, and know too what each can do, they are hardly in a position to determine which of them will best do what they want.

Our Samplers show the use to which the stitches on them may be put.

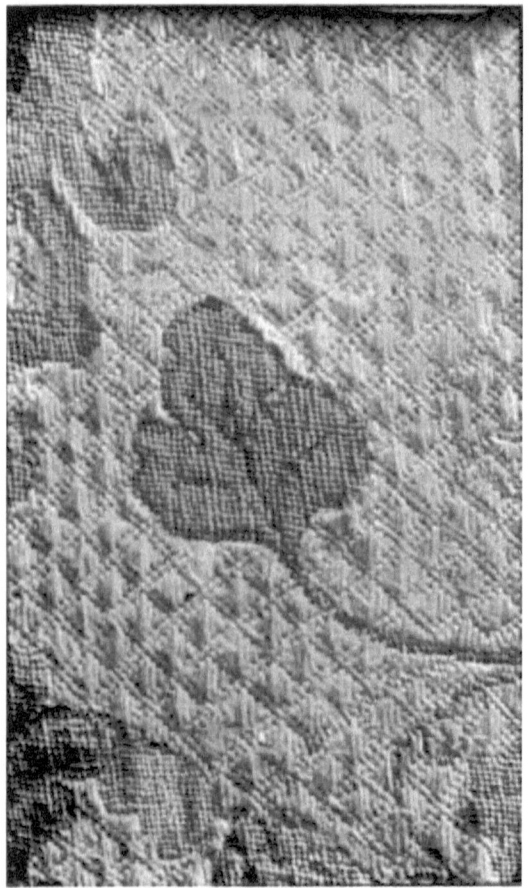

71. SATIN-STITCH IN THE MAKING.

By way of *résumé*, it may be added that for line [178] work, more or less fine, crewel, chain, back and rope stitches, and couched cord are most suitable; crewel for long lines especially, and rope stitch for both curved and straight lines; for a boundary line, buttonhole is most emphatic; for broader lines, herring-bone, feather, and Oriental stitches answer better; ladder-stitch has the advantage of a firm edge on both sides of it. Satin and chain stitches, couching and laying, and basket work make good bands, but are not peculiarly adapted to that purpose.

For covering broad surfaces, crewel, chain, and satin stitches (including, of course, what are called long-and-short and plumage stitches) serve admirably, as does also darning and laid-work; and for gold thread, couching. French knots do best for small surfaces only. The stitches most useful for purposes of shading are mentioned later on.

No sort of classification is possible until the number of stitches has been reduced to the necessary few, and all fancy stitches struck out of the list. Enquiry should also be made into the title of each stitch to the name by which it is known; and the names themselves should be brought down to a minimum.

Reduce them to the fewest any needlewoman will allow, and they are still, if not too many, more than are logically required. Some of them, too, describe not stitches, but ways of using a stitch. The term long-and-short, it has already been [179] explained (page 100), has less to do with a particular stitch than its proportion, and the term plumage-stitch refers more to the direction of the stitch than to the stitch itself. And so with other stitches. It is its oblique direction only which distinguishes stem-stitch from other short stitches of the kind. Running, again, amounts to no more than proportioning stitches to the mesh of the stuff, and taking several of them at one passing of the needle; and darning is but rows of running side by side. The term split-stitch describes no new stitch, but a particular treatment to which a crewel or a satin stitch is submitted.

The foregoing summaries of stitches are only by way of suggestion, something to set the embroidress thinking for herself. She must choose her own method; but it would help her, I think, to schedule the stitches for herself according to her own ways and wants. The most suitable stitch may not suit every one. Individual preference and individual aptitude count for something. It is not a question of what is demonstrably best, but of what best suits you.

ONE STITCH, OR MANY? [180]

The first thing to be settled with regard to the choice of stitch is whether to employ one stitch throughout, or a variety of stitches. Much will depend upon the effect desired. Good work has been done in either way; but one may safely say, in the first place, that it is as well not to introduce variety of stitch without good cause — there is safety in simplicity — and in the second, that stitches should be chosen to go together, in order that the work may look all of a piece. When the various stitches are well chosen, it is difficult at a glance to distinguish one from another.

A great variety of stitches in one piece of work is worrying, if not bewildering. It is as well not to use too many, to keep in the main to one or two, but not to be afraid of using a third, or a fourth, to do what the stitch or stitches mainly relied upon cannot do.

larger image

72. STITCHES IN COMBINATION.

It tends also towards simplicity of effect if you use your stitches with some system, not haphazard, and in subordination one to the other; there must be no quarrelling among them for superiority. You should determine, that is to say, at the outset, [182] which stitch shall be employed for filling, which for outline; or which for stalks, which for leaves, and which for flowers. Or, supposing you adopt one general stitch throughout, and introduce others, you should know why, and make up your mind to employ your second for emphasis of form, your third for contrast of texture, or for some other quite definite purpose.

It is not possible here to point out in detail the system on which the various examples illustrated have been worked; the reader must worry that out for herself. But one may just point out in passing how well the various stitches go together in some few instances.

Nothing could be more harmonious, for example, than the combination of knot, chain, and buttonhole stitches in Illustration 24; or of ladder, Oriental, herring-bone, and other stitches in Illustration 72. Again, in Illustration 85 the contrast between satin-stitch in the bird and couched cord for the clouding is most judicious, as is the knotting of the bird's crest. Laid floss contrasts, again, admirably with couched gold in Illustrations 47, 48, 49, and satin-stitch with couching in Illustration 91, where the gold is reserved mainly for outline, but on occasion serves to emphasise a detail.

larger image

73. FINE NEEDLEWORK UPON LINEN.

Couched gold and surface satin-stitch are used together again in Illustration 58, each for its specific purpose. The harmony between appliqué [184] work and couching or chain-stitch outline has been alluded to already.

A danger to be kept in view when working in one stitch only is, lest it should look like a woven textile, as it might if very evenly worked. Some kinds of embroidery seem hardly worth doing nowadays, because they suggest the loom. That may be a reason for some complexity of stitch, in which lurks that other danger of losing simplicity and breadth. The lace-like appearance of the needlework upon fine linen in Illustration 73, results chiefly from the extraordinary delicacy with which it is done, but it owes something also to the variety of stitch and of stitch-pattern employed in it.

OUTLINE. [185]

The use of outline in embroidery hardly needs pointing out. It is often the obvious way of defining a pattern, as, for example, where there is only a faint difference in depth of tint between the pattern and its background; in appliqué work it is necessary to mask the joins; and it is by itself a delightful means of diapering a surface with not too obtrusive pattern.

Allusion to the stitches suitable to outline has been made already (see stitch-groups), as well as to the colour of outlining, *à propos* of appliqué. It is difficult to overrate the importance of this question of colour in the case of outline; but there are no rules to be laid down, except that a coloured outline is nearly always preferable to a black one. The Germans of the 16th century were given to indulging in black outlines, and you may see in their work how it hardened the effect, whereas a coloured outline may define without harshness. The Spaniards, on the other hand, realised the value of colour, and would, for example, outline gold and silver upon a dark green ground in red, with admirable effect. A double outline, for which [186] there is often opportunity in bold work, may be turned to good account. Among the successful combinations which come to mind is an appliqué pattern in yellow and white upon dark green, outlined first with gold cord, and then, next the green, with a paler and brighter green. Another is a pattern chiefly in yellow upon purple, outlined first with yellow couched with gold, and next the ground with silver. In the case of couched cord or gold, the colour of the stitching counts also.

Stitches from the edge of a leaf or what not, inwards, alternately long and short, though they form an edge to the leaf, are not properly outlining. This is rather a stopping short of solid work than outlining, though it often goes by that name.

The first condition of a good outline stitch is that it should be, as it were, supple, so as to follow the flow of the form. At the same time it should be firm. Fancy stitches look fussy; and a spikey outline is worse than none at all.

There is absolutely no substantial ground for the theory that outlines should be worked in a stitch not used elsewhere in the work. On the contrary, it is a good rule not to introduce extra stitches into the work unless they give something which the stitches already employed will not give. The simplest way is always safest.

An outline affords a ready means of clearing up edges; but it should not be looked upon merely [187] as a device for the disguise of slovenliness. Unless the colour scheme should necessitate an outline, an embroidress, sure of her skill, will often prefer not to outline her work, and to get even the drawing lines within the pattern, by VOIDING. She will leave, that is to say, a line of ground-stuff clear between the petals of her flowers, or what not; which line, by the way, should be narrower than it is meant to appear, as it looks always broader than it is. It is more difficult, it must be owned, thus to work along two sides of a line of ground-stuff than to work a single line of stitching, but it is within the compass of any skilled worker; and skilled workers have delighted in voiding even when their work was on a small scale necessitating fine lines of voiding (Illustrations 39 and 40).

In work on a bold scale there is no difficulty about it; and it would be remarkable that it is so seldom used, were it not that the uncertain worker likes to have a chance of clearing up ragged edges, and that voiding implies a broader and more dignified treatment of design than it is the fashion to affect.

SHADING. [188]

One arrives inevitably at gradation of colour in embroidery; the question is how best to get it. But, before mentioning the ways in which it may be got, it seems necessary to protest that shading is not a matter of course. Perfectly beautiful work may be done, and ought more often to be done, in merely flat needlework; the gloss of the silk and its varying colour as it catches the light according to the direction of the stitching, are quite enough to prevent a monotonously flat effect.

Still, embroidery affords such scope for gradation of colour, not, practically, to be got by any process of weaving, that a colourist may well revel in the delights of colour which silks of various dyes allow. And so long as colour is the end in view there is not much danger that a colourist will go wrong.

larger image

74. PART OF A DESIGN BY WALTER CRANE.

The use of shading in embroidery is rather to get gradation of colour than relief of form. As to the stitch to be employed, that is partly a personal matter, partly a question of what is to be done. The stitch must be adapted to the kind [190] of shading, or the shading must be designed to suit the stitch. It makes all the difference in the world, whether your shading is deliberately done, or whether one shade is meant to merge into another. In the best work it is always done with decision. There is nothing vague or casual, for example, about the shading of Mr. Crane's animals on Illustration 74. Everywhere the shading is *drawn*, either in lines or as a sharply defined mass. Given a drawing in which the shadows are properly planned and crisply drawn like that, and you may use what stitch you please.

larger image

75. SHADING IN CHAIN-STITCH.

The more natural way of shading is to let the stitches follow the lines of the drawing, and so make use of them to express form, as with the strokes of the pen or pencil upon paper. Thus, in mediæval figurework prior to the 15th century, the faces were usually done in split stitch, worked concentrically from the middle of the cheek outward, and so suggesting the roundness of the face (Illustration 87). But just as there is a system of shading according to which the draughtsman makes all his strokes in one direction (slanting usually), so the embroidress may, if she prefer, take her stitches all one way; and in the 15th and 16th centuries the fashion was to work flesh in short-satin stitches always in the vertical direction (Illustration 79). The term "long-and-short-stitch" is frequently used by way of describing the stitch. It does not, as I have said, help [192] us much. The stitches are in the first place only satin-stitches worked not in even rows, as in Illustration 40, but so that there is no line of demarcation between one row and another. And this, in the case of gradated colour, makes the shading softer. The words long-and-short apply strictly only to the outer row of stitches. You begin, that is to say, with alternately long and short stitches. If you work after that with stitches of equal length, they necessarily alternate or dovetail. If the form to be worked necessitates radiation in the stitching, there results a texture something like the feathering of a bird's breast (Illustration 85), whence the name plumage-stitch, another term describing not so much a stitch as the use of a stitch.

No matter what the stitch, one must be able to draw in order to express form: it is rather more difficult to draw with a needle than with a pen, that is all. True, the designer may do that for you, and make such a workmanlike drawing that there is no mistaking it; but it takes a skilled draughtsman to do it.

larger image

76. SHADING IN SHORT STITCHES.

In flattish decorative work, where the drawing is in firm lines, as in Illustration 87, the task of the embroidress is relatively easy — there is not much shading, for example, in the drapery of King Abias, and the vine leaves are merely worked with yellower green towards the edges. Even where there is strong shading, a draughtsman [194] who knows his business may make shading easy by drawing his shadows with firm outlines. The taste of the artist who designed the roses in Illustration 75 is too pictorial to win the heart of any one with a leaning towards severity of design; too much relief is sought; but the way he has got it shows the master workman; he has deliberately laid in *flat* washes of colour, each with its precise outline, which the worker had only to follow faithfully with flat tambour work. A design like that, given the working drawing, asks little of the worker beyond patient care: of the designer it asks considerable knowledge.

A yet more pictorial effect is produced in much the same way, this time in satin stitch, in Illustration 76. The artist has for the most part drawn his shadows with crisp brush strokes, which the worker had no difficulty in following; but there is some rounding of the birds' bodies which a merely mechanical worker could not have got. In fact, there are indications that this is the work more of a painter than of an embroidress, who would have acknowledged by her stitches the feathering of the birds' necks as well as their roundness.

larger image

77. SHADING IN LONG-AND-SHORT AND SPLIT STITCHES.

You can embroider, of course, without knowing much about drawing; but you cannot go far in the direction of shading (not drawn for you, or only vaguely drawn) without the appreciation of form which comes only of knowing and understanding. There is evidence of such knowledge and understanding [196] in the working of the lion in Illustration 77. That is not a triumph of even stitching; but it is a triumph of drawing with the needle. The short satin and split stitches are not placed with the regularity so dear to the human machine, but they express the design perfectly. The embroiderer of that lion was an artist, perhaps the artist who designed it. "It might be a *man's* work," was the verdict of an embroidress. At all events it is the work of some one who could draw, and only a draughtsman or draughtswoman could have worked it.

This is not said wholly in praise of shading. Embroidery ought, for the most part, to do very well without it. The point to insist upon is that, if shading is employed at all, it should mean something, and not be mere fumbling after form.

The charm of shading in embroidery is not the roundness of form which you get, but the gradation of colour which it gives. This may be very delicately and subtly got by split-stitch, which renders that stitch so valuable in the rendering of flesh tints. But the blending of colour into colour which is universally admired is not quite so admirable as people think. One may easily employ too many shades of colour, easily merge them too imperceptibly one into the other, getting only unmeaning softness. An artist prefers to see few shades employed, and those chosen with judgment and placed with deliberate [197] intention. If they mean something, there is no harm in letting it be seen where they meet: broad masses give breadth: vagueness generally means ignorance. That is, perhaps, why one dislikes it, and why it is so common.

FIGURE EMBROIDERY. [198]

To an accomplished needlewoman embroidery offers every scope for art, short of the pictorial; and the artist is not only justified in lavishing work upon it, but often bound to do so, more especially when it comes to working with materials in themselves rich and costly. A beautiful material, if you are to better it (and if not why work upon it at all?), must be beautifully worked. Costly material is worth precious work; and there should be by rights a preciousness about the needlework employed upon it, preciousness of design and of execution. To put the value into the material is mere vulgarity.

It seems to an artist almost to go without saying, that the labour on work claiming to be art should be in excess of the value of the stuff which goes to make it. What we really prize is the hand work and the brain work of the artist; and the more precious the stuff he employs, the more strictly he is bound to make artistic use of it. I do not mean by that *pictorial* use. You can get, no doubt, with the needle effects more or less pictorial—most often less; but, when got, they are usually at the [199] best rather inferior to the picture of which they are a copy.

Work done should be better always than the design for it, which was a project only, a promise. The fulfilment should be something more. A design of which the promise is not likely to be fulfilled in the working-out is, for its purpose, ill-designed. To say that you would rather have the drawing from which it was done (and that is what you feel about "needle pictures") is most severely to condemn either the designer or the worker, or perhaps both. Only a competent figure painter, for example, can be trusted to render flesh with the needle; her success is in proportion to her skill with the implement, but in any case less than what might be achieved in painting: then why choose the needle?

Admitting that a painter who by choice or chance takes to the needle may paint with it satisfactorily enough, that does not go to prove the needle a likely tool to paint with. It is anything but that. There was never a greater mistake than to suppose, as some do who should know better, that, to raise embroidery to the rank of art, figure work is necessary. The truth is that only by rare exception does embroidered figure work rise to the rank of art: the rule is that it is degraded, the more surely as it aims at picture. And that is why, for all that has been done in the way of wonderful picture work, say by the Italians [200] and the Flemings of the Early Renaissance, the pictorial is not the form of design best suited to embroidery.

Needlework, like any other decorative craft, demands treatment in the design, and the human figure submits less humbly to the necessary modification than other forms of life. Animals, for instance, lend themselves more readily to it, and so do birds; fur and feathers are obviously translatable into stitches. Leaves and flowers accommodate themselves perhaps better still; but each is best when it is only the motive, not the model, of design. If only, then, on account of the greater difficulty in treating it, the figure is not the form of design most likely to do credit to the needle, and it is absurd to argue that, figure work being the noblest form of design, therefore the noblest form of embroidery must include it.

The embroidress entirely in sympathy with her materials will not want telling that the needle lends itself better to forms less fixed in their proportions than the human figure; the decorator will feel that there is about fine ornament a nobility of its own which stands in need of no pictorial support; the unbiassed critic will admit that figure design of any but the most severely decorative kind is really outside the scope of needle and thread; and that the desire to introduce it arises, not out of craftsmanlikeness, but out of an ambition which does not pay much regard to the conditions proper to [201] needlework. Those conditions should be a law to the needlewoman. What though she be a painter too? She is painting now with a needle. It is futile to attempt what could be better done with a brush. She should be content to work the way of the needle. Common sense asks that much at least of loyalty to the art she has chosen to adopt.

Wonderful and almost incredibly pictorial effects have been obtained with the needle; but that does not mean to say it was a wise thing to attempt them. The result may be astonishing and yet not worth the pains. The pains of flesh-painting with the needle (if not the impossibility of it for all practical purposes) is confessed by the habit which arose of actually painting the flesh in water colour upon satin. Paint on satin, if you like. There may be occasions when there is no time to stitch, and it is necessary for some ceremonial and more or less theatric purpose to paint what had better have been worked. The more frankly such work acknowledges its temporary and makeshift character the better. Scene painting is art, until you are asked to take it for landscape painting. Anyway, the mixture of painting and embroidery is not to be endured; and it is a poor-spirited embroidress who will thus confess her weakness and call on painting to help her out. It does not even do that, it fails absolutely to produce the desired effect. The painting quarrels [202] with the stitching, and there is after all no semblance of that unity which is the very essence of picture.

larger image

78. CHINESE CHAIN-STITCHING.

An instance of painted flesh occurs upon Illustration 91. Can any one, in view of the bordering to the picture, doubt that the worker had much better have kept to what she could do, and do perfectly, ornament? An example, on the other hand, of what may be done in the way of expressing action in the fewest and simplest chain stitches (if only you know the form you want to represent and can manage your needle) is given in the wee figures in the landscape above (78).

larger image

79. FIFTEENTH CENTURY FIGURE WORK.

In speaking of the necessary treatment of the human figure (as of other natural form) in needlework, it is not meant to contend that there is one [204] only way of treating it consistently, or that there are no more than two or three ways. There are various ways, some no doubt yet to be devised, but they must be the ways of the needle. The flesh, of course, is the main difficulty. A Gothic practice, and not the least happy one, was to show the flesh in the naked linen of the ground, only just working the outlines of the features in black or brown. Another way was to work the face in split stitch, as already explained, and over that the markings of the features, the fine lines in short satin-stitches, the broader in split-stitch, as shown in the figure of King Abias in Illustration 87.

The general treatment of the figure there is of course in the manner of the 14th century, better suited, from its severe simplicity, for rendering in needlework than later and more pictorial forms of composition. That needlework can, however, in capable hands, go farther than that is shown in Illustration 79, a rather threadbare specimen of 15th century work, in which the character of the man's face is admirably expressed. It is first worked in short, straight stitches, all of white, and over that the drawing lines are worked in brown. The artist gets her effect in the simplest possible way, and apparently with the greatest ease.

80. SIXTEENTH CENTURY ITALIAN FIGURE WORK.

More like painting is the head in Illustration 80, worked in short stitches of various shades, which give something of the colour as well as the [206] modelling of flesh. This is a triumph in its way. It goes about as far as the needle can go, and further than, except under rare conditions, it ought to go. But it may do that and yet be needlework.

Equally wonderful in their miniature way are the faces of the little people on Illustration 81, about the size of your finger nail. They are worked in solid satin-stitch, and the two layers of silk (back and front) give a substance fairly thick but at the same time yielding, so that when the stitches for the mouth and eyes are sewn tightly over it they sink in, and, as it were, push up the floss between and give relief. The nose is worked in extra satin-stitch over the other, and the slight depression at the end of the stitch gives lines of drawing. This trenches upon modelling, but, on such a minute scale, does not amount to very pronounced departure from the flat. The method employed does not lend itself to larger work.

The last word on the question as to what one may do with the needle is, that you may do what you *can*; but it is best to seek by means of it what it can best do, and always to make much of the texture of silk, and of the quality of pure and lustrous colour which it gives—in short, to work *with* your materials.

larger image
81. CHINESE FIGURES.

THE DIRECTION OF THE STITCH. [208]

The effect of any stitch is vastly varied, according to the use made of it. Satin-stitch, it was shown (38), worked in twisted silk, ceases to have any appearance of satin; and it makes all the difference whether the stitches are long or short, close together or wide apart. More important than all is the direction of the stitch. By that alone you can recognise the artist in needlework.

The DIRECTION of the stitch deserves consideration from two points of view — that of colour and that of form. First as to colour. It is not sufficiently realised that every alteration in the direction of the stitch means variety of tone, if not of tint. Take a feather in your hand, and turn it about, so that now one side of the quill now the other catches the light; or notice the alternate stripes of brighter and greyer green on a fresh-trimmed lawn, where the roller has bent the blades of grass first this way and then that. So it is with the colour of silken stitches. The pattern opposite (82) looks as if it had been embroidered in two shades of silk; in the work itself it has still more that appearance; but [210] it is all in one shade of brownish gold: the difference which you see is merely the effect of light upon it. The horizontal stitches, as it happens, catch the light; the vertical ones do not. Had the light come from a different point, the effect might have been reversed. If there had been diagonal stitches from right to left, they would have given a third tint; and, if there had been others from left to right, they would have given a fourth.

larger image

82. INFLUENCE OF STITCH-DIRECTION UPON COLOUR.

Suppose a pattern in which the leaves were worked horizontally, the flowers vertically, and the stalks in the direction of their growth, all in one stitch and in one colour, there would be a very appreciable difference in tone between leaves, flowers, and stalks. In gold, the difference would be yet more striking. And that is one reason why gold backgrounds are worked in diapers; not so much for the sake of pattern as to get variety of broken tint.

In the famous Syon Cope the direction of the stitching is frankly independent of the design. That is to say, that, while the pattern radiates naturally from the neck, the stitches do not follow suit, but go all one way—the way of the stuff. This, though rather a brutal solution of the difficulty, saves all afterthought as to what direction the stitches shall take; but it has very much the effect of weaving. The embroiderer of the 13th century was not afraid of that (aimed at it, [211] perhaps?), and was, apparently, afraid of letting go the leading strings of warp and weft.

When stitches follow the direction of the form embroidered, accommodating themselves to it, all manner of subtle change of tone results. You get, not only variety of colour, but more than a suggestion of form.

That is the second point to be considered.

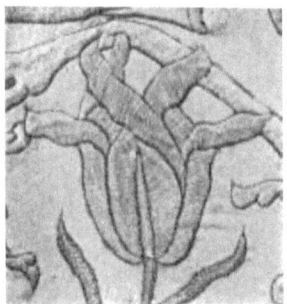

larger image

83. MEANINGLESS DIRECTION OF STITCH.

The direction taken by the stitch always helps to explain the drawing; or, if the needlewoman cannot draw, to show that she cannot—as, for example, in the tulip herewith (83). A less intelligent management of the stitch it would be hard to find. The needlestrokes, far from helping in the very slightest degree to explain the folding over of the petals, directly contradict the drawing. The flower might almost have been designed to show how not to do it; but it is a piece of old work, quite seriously done, only without knowing. The embroidress is free, of course, to work her stitches in a direction which does not express [212] form at all, so as to give a flat tint, in which is no hint of modelling; but the intention is here quite obviously naturalistic. The rendering below (84) shows the direction the stitches should have taken. The turn-over of the petals is even there not very clearly expressed, but that is the fault of the drawing (very much on a par with the workmanship), from which it would not have been fair to depart.

larger image

84. MORE EXPRESSIVE LINES OF STITCHING.

A more clever fulfilment of the naturalistic intention is to be seen in Illustration 76. The drawing of the doves is in the rather loose manner of the period of Marie Antoinette; but the treatment of the stitch is clever in its way — the way, as I have said, rather of painting than of embroidery, giving as it does the roundness of the birds' bodies but no hint of actual feathering, such as you find in the bird in Illustration 85. There, every stitch helps to explain the feathering. By a discreet use of what I must persist in calling the same stitch (that is, satin-stitch and the variety of it called [214] plumage-stitch) the embroiderer has rendered with equal perfection the sweep of the broad wing feathers and the fluffy feathering of the breast. It is by means of the direction of the stitch, too, that the drawing of the neck is so perfectly rendered.

larger image

85. SATIN AND PLUMAGE STITCHES.

The direction of the stitch is varied to some purpose in the head in Illustration 80, where the flesh is all in straight upright stitches, whilst the hair is stitched in the direction of its growth.

The five petals on the satin-stitch sampler (Illustration 36) — to descend from the masterly to the elementary — show something of the difference it makes in what direction the stitch is worked. It matters more, of course, in some stitches than in others; but in most cases the direction of the stitch suggests form, and needs accordingly to be considered.

It scarcely needs further pointing out how the direction of the stitch may help to explain the construction of the form, as in the case of leaves, for example, where the veining may be suggested; or of stalks, where the fibre may be indicated. There is no law as to the direction of stitch, except that it should be considered. You may follow the direction of the forms, you may cross them, you may deliberately lay your stitches in the most arbitrary manner; but, whatever you do, you must do it with intelligent purpose. An artist or a workwoman can tell at once whether your stitch was [215] laid just so because you meant it or because you knew no better.

Having laid your stitches deliberately, it is best to leave them, and not to work over them with other stitching. Stitching over stitching was resorted to whenever elaboration was the fashion; but the simpler and more direct method is the best. The way the veins are laid in cord over the satin-stitch in the lotus leaves in Illustration 40 is the one fault to be found with an all but perfect piece of work.

The stitching over the laid silver mid-rib in Illustration 92 is better judged. It may be said, generally speaking, that except where, as in the case of laid-work, the first stitching was done in anticipation of a second, and the work would be incomplete without it, stitching over stitches should be indulged in only with moderation.

Stitching is sometimes done not merely over stitches, but upon the surface of them, not penetrating the ground-stuff. Unless, in such a case, the first stitching is of such compact character as to want no strengthening, it amounts almost to a sin against practicality not to take advantage of the second stitching to make it firmer.

CHURCH WORK. [216]

It is customary to draw a distinction between church, or ecclesiastical as it is called, and other embroidery; but it is a distinction without much difference. Certain kinds of work are doubtless best suited to the dignity of church ceremonial, and to the breadth of architectural decoration; accordingly, certain processes of work have been adopted for church purposes, and are taken as a matter of course—too much as a matter of course. The fact is, work precisely like that employed on vestments and the like (Illustration 86) was used also for the caparison of horses and other equally profane purposes.

larger image

86. RENAISSANCE CHURCH WORK.

Practical considerations, alike of ceremonial and decoration, make it imperative that church work should be effective: religious sentiment insists that it should be of the best and richest, unsparingly, and even lavishly given; common sense dictates that the loving labour spent upon it should not be lost. And these and other such considerations involve methods of work which, by constant use for church purposes, have come to be classed as ecclesiastical embroidery. But there is no [218] consecrated stitch, no stitch exclusively belonging to the church, none probably invented by it. For embroidery is a primitive art—clothes were stitched before ever churches were furnished; and European methods of embroidery are all derived from Oriental work, which found its way westwards at a very early date. Phrygia (sometimes credited with the invention of embroidery) passed it on to Greece, and Greece to Italy, the gate of European art.

Christianity produced new forms of design, but not new ways of work. The methods adopted in the nunneries of the West were those which had already been perfected in the harems of the East.

Embroidery for the church must naturally take count of the church, both as a building and as a place of worship; but, as apart from all other needlework, there is no such thing as church embroidery; and the branding of one very dull kind of thing with that name is in the interest neither of art nor of the church, but only of business. "Ecclesiastical art" is just a trade-term, covering a vast amount of soulless work. There is in the nature of things no reason why art should be reserved for secular purposes, and only manufacture be encouraged by the clergy. The test of fitness for religious service is religious feeling; but that is hardly more likely to be found in the output of the church furnisher (trade patterns overladen with stock symbols), than in [219] the stitching of the devout needlewoman, working for the glory of God, in whose service of old the best work was done.

Many of the examples of old work given on these pages are from church vestments, altar furniture, and the like; information on that point will be found in the descriptive index of illustrations at the beginning of the book; but they are here discussed from the point of view of workmanship, with as little reference as possible to religious or other use: that is a question apart from art.

The distinguishing features of church work should be, in the first place, its devotional spirit, and, in the second, its consummate workmanship. In it, indeed, we might expect to find work beyond the rivalry of trade controlled by conditions of time and money. Even then it would be but the more perfect expression of the same art which in its degree ennobled things of civic and domestic use.

Church embroidery, as usually practised in these days, is not only the most frigid and rigid in design, but the hardest and most mechanical in execution—which last arises in great part from the way it is done. It is not embroidered straight upon the silk or velvet which forms the groundwork of the design, but separately on linen. The pattern thus worked is cut out, and either pasted straight on to the ground-stuff, or, if the linen is at all loose, first mounted on thin paper and then cut out and [220] pasted on to the velvet, where it is kept under pressure until it is dry. In either case the edges have eventually to be worked over.

This habit of working on linen or canvas and applying the embroidery ready worked on to the richer stuff, though early used on occasion, does not seem to have been common until a period when manufacture generally usurped the place of art. The work in Illustration 87 was done directly on to the silk. In the latter half of the 18th century there was a regular trade in embroidery ready to sew on, by which means purveyors could turn out in a day or two what would have taken months to embroider.

Even if it had been the invariable mediæval practice to work sprays or what not upon canvas and apply them bodily to the velvet, that would not make it the more workmanlike or straightforward way of working. If needle stitches are the ostensible means of getting an effect upon a stuff, it seems only right they should be stitched upon that stuff. To work the details apart and then clap them on to it, stands to embroidery very much in the relation of hedge-carpentering to joinery. Nor is it usually happy in result. Occasionally, as in the case of Miss C. P. Shrewsbury's vine-leaf pattern (Illustration 88), it disarms criticism. More often it looks stuck-on. A way of avoiding that look is to add judicious after-stitching on the stuff itself; and this must not be confined to the [222] sewing on or outlining merely, but allowed to wander playfully over the field, so as to draw your eye away from the margin of the applied patch, and lead you to infer that, some of the needlework being obviously done on the velvet, all of it is. But to disguise in this way the line of demarcation, even if you succeed in doing it, is at best the art of prevarication.

larger image

87. GOTHIC CHURCH WORK.

No doubt it is difficult to work upon velvet. The stuff is not very sympathetic, and the stitching has a way of sinking into the pile, and being, as it were, drowned in it. But the trailing spirals of split-stitch which play about the applied spots in many a mediæval altar cloth hold their own quite well enough to show that silk can be worked straight on to the velvet.

That gold may be equally well worked straight on to velvet may be seen in any Indian saddle cloth. Heavy work of this kind may be rather man's work than woman's; but that is not the point. The question is, how to get the best results; and the answer is, by working on the stuff.

It may be argued that in this way you cannot get very high relief; but the occasions for high relief are, at the best, rare. If you want actual modelling, as in the Spanish work referred to in a previous chapter, that must, of course, be worked separately, built up, as it were, upon the canvas and worked over. And there is no reason why it [224] should not, for in no case does it appear to be stitching. In fact, it aims deliberately at the effect of chased and beaten metal.

larger image

88. MODERN CHURCH WORK BY MISS SHREWSBURY.

Heavy appliqué of any kind affects, of course, not only the thickness but the flexibility of the material thus enriched — an important consideration if it is meant to hang in folds.

A PLEA FOR SIMPLICITY. [225]

The simplest patterns are by no means the least beautiful. It is too much the fashion to underrate the artistic value of the less pretentious forms of needlework, and especially of flat ornament, which has, nevertheless, its own very important place in decoration. As for geometric pattern, that is quite beneath consideration—it is so mechanical! Mechanical is a word as easily spoken as another; but if needlework is mechanical, that is more often the fault of the needlewoman than of the mechanism she employs. The Orientals, who indulged so freely in geometric device, were the least mechanical of workers. It is our rigid way of working it which robs geometric ornament of its charm. The needleworker has less than ever occasion to be afraid of geometric pattern; for it is peculiarly difficult to get in it that appearance of rule-and-compass-work which makes ornament so dull.

The one real objection to geometric pattern is that it is nowadays so cheaply and so mechanically got by *weaving* that, however freely it may be rendered, there is a danger of its suggesting [226] mechanical production, which embroidery emphatically ought not to do. There is a similar objection nowadays to some stitches, such, for example, as chain-stitch and back-stitch, which suggest the sewing-machine.

Embroidery does not to-day take quite the place it once did. It was used, for example, by the early Coptic Christians to supplement tapestry. That is to say, what they could not weave they stitched; it was only to get more delicate detail than their tapestry loom would allow, that they had recourse to the needle. Needlework was, in fact, an adjunct to weaving. Later, in mediæval times, the Germans of Cologne, for their church vestments and the like, wove what they could, and enriched their woven figures with embroidery.

Again, a great deal of Oriental embroidery, and of peasant work everywhere, is merely the result of circumstances. Where money is scarce and time is of no account, it answers a woman's purpose to do for herself with her needle what might in some respects be even better done on the loom. Her preference for handwork is not that it has artistic possibilities, but that it costs her less. She would in many cases prefer the more mechanically produced fabric, if she could get it at the same price. We do not find that Orientals reject the productions of the power-loom—which they would do if they had the artistic instincts with which we credit them.

larger image

89. SIMPLE STITCHING ON LINEN.

[228] It results from our conditions of to-day that there are some kinds of needlework we admire, which yet are not worth our doing, such, for example, as the all-over work, which does not amount to more than simple diaper, and which really is not so much embroidering on a textile as converting it into one of another kind. Glori-

fied instances of this kind of work occur in the shawl work of Cashmere, and in those beautiful bits of Persian stitching which remind one of carpet-work in miniature, if they are not in fact related to carpet-weaving.

Embroidery was at one time the readiest, and practically the only, means of getting enrichment of certain kinds. To-day we get machine embroidery. As machinery is perfected, and learns to do what formerly could be done only by the needle, hand-workers get pushed aside and fall out of work. Their chance is, in keeping always in advance of the machine. There is this hope for them, that the monotony of machine-made things produces in the end a reaction in favour of handwork — provided always it gives us something which manufacture cannot. Possibly also there is scope for amateurs and home-artists in that combination of embroidery and hand-weaving with which the power-loom, though it has superseded it, does not enter into competition.

larger image

90. SIMPLE COUCHING ON LINEN.

It is not so much for geometric ornament as for simple pattern that I here make my plea, for [230] that reticent work of which so much was at one time done in this country—mere back-stitching, for example, or what looks like it, in yellow silk upon white linen; or the modest diaper, archaic, if you like, but inevitably characteristic, in which the naïveté of the sampler seems always to linger; or again, the admirably simple work in Illustration 89. This last does not show so delicately in the photographic reproduction as it should, because, being in grey and yellow on white linen, the relative value of the two shades of colour is lost in the process. In the original the broader yellow bands are much more in tone with the ground, and do not assert themselves so much. Such as it is, only an artist could have designed that border-work, and any neat-handed woman could have embroidered it.

Think again of the delicate work in white on white, too familiar to need illustration, which makes no loud claim to be art, but is content to be beautiful! Is that to be a thing altogether of the past now that we have Art Needlework? Art needlework! It has helped put an end to the patience of the modern worker, and to inspire her too often with ambitions quite beyond her powers of fulfilment.

What one misses in the work of the present day is that reticent and unpretending stitchery, which, thinking to be no more than a labour of loving patience, is really a work of art, better deserving [231] the title than a flaunting floral quilt which goes by the name of "art needlework"—designed apparently to worry the eye by day and to give bad dreams by night to whoever may have the misfortune to sleep under it. Is anyone nowadays modest enough to do work such as the couching in outline in Illustration 90? Yet what distinction there is about it!

EMBROIDERY DESIGN. [232]

Perfect art results only when designer and worker are entirely in sympathy, when the designer knows quite what the worker can do with her materials, and when the worker not only understands what the designer meant, but feels with him. And it is the test of a practical designer that he not only knows the conditions under which his design is to be carried out, but is ready to submit to them.

The distinction here made between designer and embroiderer is not casual, but afore-thought, notwithstanding the division of labour it implies. Enthusiasm has a habit of outrunning reason. Because in some branches of industry subdivision of labour has been carried to absurd excess, it is the fashion to demand in all branches of it the autograph work of one person, which is no less absurd. To try and link together faculties which Nature has for the most part put asunder, is futile.

That designer and worker should be one and the same person is an ideal, but one only very occasionally fulfilled. When that happens (Illustrations 61 [233] and 88) it is well. But the attempt to realise it commonly works out in one of two ways: either a good design is spoilt in the working for want of executive skill on the part of the designer, or good workmanship is spent on poor design, as good, perhaps, as one has any right to expect of a skilled needleworker.

The fact is, you can only make out all the world to be designers by reducing design to what all the world can do. And that is not much. There is a point of view from which it does not amount to design at all.

The study of design forms part of the education of an embroidress, not so much that she may design what she works, but that she may know in the first place what good design is, and, in the second, be equal to the ever-recurring occasion when a design has to be modified or adapted. If, in thus manipulating design not hers, she should discover a faculty of invention, she will want no telling to exercise it. A designer wants no encouragement to design—she designs.

There would be no occasion to insist upon this, were it not for the prevalence at the present moment of the idea that a worker, in whatever art or handicraft, is in artistic duty bound to design whatever she puts hand to do. That is a theory as false as it is unkind; let no embroidress be discouraged by it. Let her, unless she is inwardly [234] impelled to invent, remain content to do good needlework. That is her art. Her business as an artist is to make beautiful things. Co-operation in the making of them is no crime.

And what, then, about originality? Originality is a gift beyond price. But it is not a thing which even the designer should struggle after. It comes, if it is there. There is a revengeful consolation for the pain we suffer from design about us writhing to be up-to-date, in the thought that its contortions tell what pain it cost to do. The birth of beauty is a less agonising travail; and the thing to seek is beauty, not novelty. Whoever planned the lines of the border in Illustration 91, or treated the leafage in Illustration 92, was not trying to be original, but determined to do his best. Artists and workers of individuality and character are themselves, without being so much as aware that originality has gone out of them.

larger image

91. RENAISSANCE ORNAMENT.

To assume, then, that every needlewoman is, or can ever be, competent to design what she embroiders, is to make very small account of design. How is it possible to take design seriously and yet think it is to be mastered without years of patient study, which few workwomen can or will devote to it? Any cultivated woman may for herself invent (if it is to be called invention) something better worth working than is to be bought ready to work. And that may do for many purposes, so long as it does not claim to be [236] more than it is; but in the case of really important work, to be executed at considerable cost not only of material, but of patient labour, surely it is worth giving serious thought to its design. The scant consideration commonly given to it shows how little the worker is in earnest. Or has she thought? And is she persuaded that her artless spray of flowers, or the ironed-off pattern she has bought, is all that art could be? It would be rude to tell her she was wasting silk! How should she know?

The only way of knowing is to study, to look at good work, old work by preference; it is worth no one's while to praise that unduly. And if in all that is now so readily accessible she finds nothing to admire, nothing which appeals to her, nothing which inspires her, then her case is hopeless. If, on the other hand, she finds only so much as one style of work sympathetic to her, studies that, lets its spirit sink into her, tries to do something worthy of it, then she is on the right road. Measure yourself with the best, not with the common run of work; and if that should put you out of conceit with your own work, no great harm is done; sooner or later you have got to come to a modest opinion of yourself, if ever you are to do even moderate things.

larger image

92. LEAF TREATMENT IN APPLIQUÉ.

But the "best" above referred to does not necessarily mean the most masterly. The best of a simple kind is not calculated to discourage [238] anyone—rather, it looks as if it must be easy to do that; and in trying to do it you learn how much goes to the doing it. Good design need not be of any great importance or pretensions. It may be quite simple, if only it is right; if the lines are true, the colour harmonious; if it is adapted to its place, to its use and purpose, to execution not only with the needle but in the particular kind of needlework to be employed.

There has of late years been something of a revival of needlework design in schools of art, and some very promising and even most accomplished work has been done; but in many instances, as it seems to me, it is rather design which has been translated into needlework, than design clearly made for execution with the needle. A really appropriate and practical design for embroidery should be schemed not merely with a view to its execution with the needle, but with a view to its execution in a particular stitch or stitches — and possibly by a particular embroidress. To be safe in designing work so minute as that on Illustration 93, one must be sure of the needlewoman who is to execute it.

larger image

93. DELICATE SATIN-STITCH—WORKED BY MISS BUCKLE.

My reference to old work must not be taken to imply that design should be in imitation of what has been done, or that it should follow on those lines. Design was once upon a time traditional; but the chain of tradition has snapped, and now conscious design must be eclectic—that is to say, [240] one must study old work to see what has been done, and how it has been done, and then do one's own in one's own way. It is at least as foolish to break quite away from what has been done as to tether yourself to it. And in what has been done you will see, not only what is worth doing, but what is not. That, each must judge for herself. For my part, it seems to me the thing best worth doing is ornament. Any way, this much is certain (and you have only to go to a museum to prove it), that there is no need for needleworkers, unless their instinct draws them that way, to take to needle painting, to pictures in silk, or even to flower stitching.

The limitations of embroidery are not so rigidly marked as the boundaries of many another craft. There is little technical difficulty in representing flowers, for example, very naturally—too naturally for any dignified decorative purpose. Embroiderer or embroidery designer will, as a matter of fact, be constantly inspired by flower forms, and silk gives the pure colour of their petals as nearly as may be. But, though the pattern be a veritable flower garden, the embroidress will not forget, to use the happy phrase of William Morris, that she is gardening with silks and gold threads.

Let the needleworker study the work of the needle in preference to that of the brush; let her aim at what stuff and threads will give her, and [241] give more readily than would something else. Let her work according to the needle: take that for her guide, not be misled by what some other tool can do better; do what the needle can do best, and be content with that. That is the way to Art in Needlework, and the surest way.

EMBROIDERY MATERIALS. [242]

Embroidery is not among the things which have to be done, and must be done, therefore, as best one can do them. It is in the nature of a superfluity: the excuse for it is that it is beautiful. It is not worth doing unless it is done well, and in material worth the work done on it. If you are going to spend the time you must spend to do good work, it is worth while using good stuff, foolish to use anything else. The stuff need not be costly, but it should be the best of its kind; and it should be chosen with reference to the work to be done on it, and *vice versâ*. A mean ground-stuff suggests, if it does not necessitate, its being embroidered all over, ground-work as well as pattern; a worthier one, that it should not be hidden altogether from view; a really beautiful one, that enough of it should be left bare of ornament that its quality may be appreciated.

STUFFS.

It goes without saying, that for big, bold stitching a proportionately coarse ground-stuff should be used, and for delicate work, one of finer texture; whether it be linen, woollen cloth, or silk, your purpose will determine. [243]

Linen is a worthy ground-stuff, which may be worked on with flax thread, crewel, or silk, but they should not be mixed. Cotton is hardly worth embroidering. Of woollen stuffs, good plain cloth is an excellent ground for work in wool or silk, but it is not pleasant to the touch in working. Serge, if not too loose, may serve for curtains and the like, but it is not so well worth working upon. Felt is beneath contempt.

The nobler the material, the more essential it is that it should be of the best. Poor satin is not "good enough to work on;" it looks poorer than ever when it is embroidered.

Satin should be stretched upon the frame the way of the stuff, and it should not be forgotten that it has a right and a wrong way up. If it is backed, the linen should be fine and smooth: on a coarse backing, the satin gets quickly worn away, as you may see in many a piece of old work that has gone ragged.

"Roman satin" and what is called "*satin de luxe*" (perhaps because it is not so luxurious as it pretends to be) are effective ground-stuffs easy to work upon; but there is an odour of pretence about satin-faced cotton.

A corded silk is not good to embroider; the work on it looks hard; but a close twill answers very well. Silk damask makes an admirable ground beautifully broken in colour, if only it is simple and broad enough in pattern. Generally speaking, [244] you can hardly choose a design too big and flat; but something depends upon the work to be done on it. In any case, the pattern of the damask ought not to assert itself, and if you can't make out its details, so much the better.

Brocade asserts itself too much to form a good background. There is a practice of embroidering the outlines, or certain details only, of damask and brocade patterns. That is a fair way of further enriching a rich stuff; but it is embroidery merely in the sense that it is literally embroidered: the needlework is only supplementary to weaving.

Tussah silk of the finer sort is easy to work in the hand. The thinner and looser quality needs to be worked in a frame, and with smooth silk not tightly twisted.

THREAD.

With regard to the thread to work with: The coarser kinds of flax are best waxed before using. The crewel to be preferred is that not too tightly twisted. Filoselle is well adapted to couching, and may be laid double (24 threads). French floss is smooth, and does well for laid work; for fine work bobbin floss, or what is called "church floss," is better; the slight twist in filo-floss is against it; very thick floss may be used for French knots.

For couching gold, a very fine twisted silk does well. Purse silk, thick and twisted, lends itself perfectly to basket work. Working in coloured silks, one should take advantage of the quality [245] of pure transparent colour which silk takes in the dyeing. The palette of the embroiderer in silk is superlatively rich.

GOLD.

The purest gold is generally made on a foundation of *red* silk. Japanese gold does not tarnish so readily as "passing," which is in some respects superior to it. For stitching through, there is a finer thread, called "tambour." Flat gold wire is known by the name of "plate," and various twisted threads by the name of "purl."

CHENILLE.

A not very promising substance to embroider with is chenille. It came into use in the latter half of the 17th century, and was still in fashion in the time of Marie Antoinette. The use of it is shown in Illustration 75, where the darker touches of the roses are worked in it. Chenille seems to have been used instead of smooth silk, much as in certain old-fashioned water-colour paintings gum was used with the paint, or over it, to deepen the shadows. The material is used again in the wreath on Illustration 76. It is worked there in chain-stitch with the tambour needle: it may also be worked in satin-stitch; but the more obvious way of using it is to couch it, cord by cord, with fine silk thread. There is this against chenille, that its texture is not sympathetic to the touch, and that there is a stuffy look about it always. Nor does it seem ever quite to belong to the smooth satin ground on which it is worked. [246]

RIBBON.

SHADED SILK.

There is less objection to embroidery in ribbon, which also had its day in the 18th century. It was very much the fashion for court dresses under Louis Seize—"*Broderie de faveur,*" as it was called, whence our "lady's favour"—*faveur* being a narrow ribbon. Some beautiful work of its kind was done in ribbon, sometimes *shaded.* Shaded silk, by the way, may be used to artistic purpose. There is, for example, in the treasury of Seville Cathedral a piece of work on velvet, 13th century, it is said, rather Persian in character, in which the forms of certain nondescript animals are at first sight puzzlingly prismatic in colour. They turn out to be roughly worked in short stitches of parti-coloured silk thread. The result is not altogether beautiful, but it is extremely suggestive.

RIBBON.

The effect of ribbon work is happiest when it is not sewn through the stuff after the manner of satin stitch, but lies on the surface of the satin ground, and is only just caught down at the ends of the loops which go to make leaves and petals. The twist of the ribbon where it turns gives interest to the surface of the embroidery, which is always more or less in relief upon the stuff, easy to crush, and of limited use therefore.

larger image
94. LEATHER APPLIQUÉ UPON VELVET.

An effect of ribbon work, but of a harder kind, was produced by onlaying narrow strips of card or parchment upon a silken ground, twisted about after the fashion of ribbon. These, having been [248] stitched in place, were worked over in satin-stitch. The work has the merit of looking just like what it is. But neither it nor ribbon embroidery is of any very serious account.

Passing reference has been made to other materials to embroider with than thread. Gold wire, for example, and spangles, coral and pearls, which have been used with admirable discretion, as well as to vulgar purpose. Jewels also were lavished upon the embroidery of bishops' mitres, gloves and other significant apparel, and in default of real stones, imitations in glass, and eventually beads (or pearls) of glass, in which we have possibly the origin of knots. Bead embroidery is at least as old as ancient Egypt. Even atoms of looking-glass, sewn round with silk, have been used to really beautiful effect (barbaric though it may be) in Indian work. The question almost occurs: with what can one not embroider? In Madras they produce most brilliant embroidery upon muslin with the cases of beetles' wings. In the Mauritius they use fish-scales; in North America, porcupine quills; and everywhere savage tribes use seeds, shells, feathers, and the teeth and claws of animals.

To return to more civilised work, there is embroidery in gold and silver wire, allied to the art of the goldsmith, and on leather (Illustration 94), allied to the art of the saddler. It would be difficult to set any limit to the directions in which [249] embroidery may branch out, impossible to describe them all. Happily, it is not necessary. A skilled worker adapts herself to new conditions, and the conditions themselves dictate the necessary modification of the familiar way.

A WORD TO THE WORKER. [250]

A good workwoman will not encumber herself with too many tools; but she will not shirk the expense of necessary implements, the simplest by preference, and the best that are made.

NEEDLES.

Embroidery needles should have large eyes; the silk is not rubbed in threading them, and they make way for the thread to pass smoothly through the stuff. For working in twisted silk, the eye should be roundish; for flat silk, long; for surface stitching or interlacing, a blunt "tapestry needle" is best; for carrying cord or gold thread through the stuff, a "rug needle."

THIMBLE.

For a thimble, choose an old one that has been worn quite smooth.

SCISSORS.

For scissors, be sure and have a strong, short, sharp and pointed pair — the surgical instrument, not the fancy article. Nail scissors would not be amiss but for the roughness of the file on the blades.

PINS.

For pins, use always steel ones; and for tacks, those which have been tinned; or they will leave their mark behind them.

FRAMES.

For a frame, get the best you can afford; a cheap [251] one is no economy; but a stand for it is not always necessary. It should be rather wider than might seem necessary, as the work should never extend to the full width of the webbing. A tambour frame is also useful, though you have no intention of doing tambour work.

TO STRETCH SILK.

In stretching silk (not backed with linen) upon a frame, some preliminary care is necessary. The stuff should first be bordered with strips of linen or strong tape, and into the two sides of this border which are to be laced up a stout string should be tacked, to prevent it from giving when the work is drawn tight.

FRAMING.

The way to put embroidery material (thus bordered or not) into a frame is: first to sew it to the webbing (top and bottom), then to put the laths or screws into the bars, tightening them evenly, and lastly to lace it to the sides with fine string and a packing needle.

TRANSFERRING.

The ordinary ways of transferring a design to embroidery material are well known: the outline may be traced down with a point over transfer paper; it may be pricked upon paper and pounced upon the stuff in chalk or charcoal, and then traced in with a brush or pen; or it (still the outline only) may be stencilled. In any case, the outline marked upon the stuff should be well within what is to be the actual outline of the embroidery when worked. Another way, more peculiarly adapted to needlework, is to trace the outline in ink upon fine [252] tarlatan (leno muslin will do for very coarse work), and, having laid this down upon the stuff, to go over the lines again with a ruling pen and Indian ink or colour. On a light stuff it is possible to use, instead of a pen, a hard pencil. On a dark material one must use Chinese white, to which it is well to add, not only a little gum (arabic), but a trace of ox-gall, to make it work easily. One gets by this method naturally rather a rotten line upon the ground-stuff, but it is enough for all practical purposes.

KEEPING CLEAN.

Delicate work is easily rubbed and soiled in the working. It is only reasonable precaution to protect it by a veil or covering of thin, soft, white glazed lining, tacked round the edges on to the stuff. On this you mark the four lines inclosing the actual embroidery, and, cutting through three of them, you have a flap of lining, which you raise and turn back when you are at work. If the work is very delicate, you may make instead of one flap a succession of little ones; but you see then only a portion of your work at a time, and cannot so well judge its effect.

STARTING AND FINISHING.

In starting work, do not begin by making a knot in your thread; run a few stitches (presently to be worked over) on the right side of the stuff. In finishing, you run them at the back of the stuff; for greater security still, one may end with a buttonhole-stitch.

PUCKERING.

There is less danger of puckering the stuff if [253] you hold it over two fingers (at least), keeping it taut and the thread loose.

Working without a frame, it often comes handiest to hold the stuff askew, and there is a natural inclination to pull it in that direction. This temptation must be resisted, or puckering is sure to result.

DOUBLE THREAD.

In working with double silk or wool, it is better not to double back a single thread, but to pass two separate threads through the eye of the needle. The four threads (where these are turned back near the eye) make way through the stuff for the double thread, which passes easily; moreover, the thread by this means is not pulled too tight, and the effect is richer.

The stitch wants always adaptation to the work it has to do. In working a curved line, for example, say in herring-bone-stitch, one is bound always to take up a larger piece of stuff on its outside than on its inner edge.

When a thread runs short, it is better not to go on working with it, but to take another; and in finishing off, remember to run the thread in the direction opposite to that from which you are going to run the new one. In starting the new stitch, you naturally bring your needle out as if it were a continuation of that last made.

UNDOING.

If your work is faulty, cut it out and do it again. Unpicking is not so satisfactory: it loosens the stuff to drag the thread back through it, and the thread [254] saved is of no further use. Beginners find it hard to undo work once done; but a really good needlewoman never hesitates about it—her one thought is to get the thing right. Don't break your thread ever: that pulls it out of condition: cut it always.

In working, it is well to keep strictly to the stitch you have chosen, but not to the point of bigotry. One may finish off darning, for example, at the edges with a satin stitch. The thing to avoid is fudging. Moreover, stitches should be laid right at once; there should be no boggling and botching, no working-over with stitches to make good — that is not playing fair.

SMOOTHING.

When the needlework is done, do not finish it with a flat iron. That finishes it in more senses than one. But suppose it is puckered? In that case, stretch it and damp it. To do this, first tack on to it (as explained on page 251) a frame of strong tape. Then, on a drawing-board or other even wooden surface, lay a piece of clean calico, and on that, face downwards, the embroidery, and, slightly stretching it, nail it down by the tape with tin-tacks rather close together. If now you lay upon it a damp cloth, the embroidery will absorb the moisture from it, and when that is removed, should dry as flat as it is possible to get it.

A rather more daring plan is to damp the back of the stuff with a wet sponge. The work, instead of being nailed on to a board, may just as well be [255] laced to a frame by the tape. In the case of raised embroidery there must be between it and the wood, not a cloth merely, but a layer of wadding.

The damping above described may take the form of a thin paste or stiffening, but upon silk or other such material this wants tenderly doing.

One last word as to thoroughness in needlework. Those who have really not time to do much, should be satisfied with simple work. The desire to make a great show with little work is a snare. Ladies make protest always, "There is too much work in that." Well, if they are not prepared to work, they may as well give themselves up to their play. There was no labour shirked in the old work illustrated in these pages; and nothing much worth doing was ever done without work, hard work, and plenty of it. Should that thought frighten folk away, they may as well be scared off at once. Art can do very well without them.

INDEX. [257]

- Adaptation of stitch, 103, 188, 253
- Antique stitch, 66 (*See also Oriental-stitch*)
- Appliqué, 140, 144 *et seq.*, 220, 222, 224
- Arab work, 152
- Artless art, 37, 236
- Attachment of cord, 124

- Backstitch, 30, 37, 41, 53, 83, 86, 172, 226, 230
- Basket patterns, 134
- Beads, 248
- Beginning & finishing, 252
- Blanket-stitch, 56
- Braid-stitch, 42, 43
- Broad surfaces (covering), 178
- Brocade, 244
- Bullion, 165
- Bullion-stitch, 75, 76, 162, 165
- Buttonhole-stitch, 8, 55 *et seq.*, 69, 122, 145, 158, 178, 182
- Buttonholing (lace), 84, 86
- Byzantine embroidery, 12, 24

- Cable-chain, 42
- Canvas, 7, 25
- Canvas stitches, 12 *et seq.*
- Canvas-stitch embroidery, 22
- Card underlay, 162, 246
- Cashmere embroidery, 228
- Cashmere-stitch, 18
- Chain-stitch, 38 *et seq.*, 61, 83, 129, 145, 156, 158, 178, 182, 202, 226, 245
- Chenille, 245
- Chinese embroidery, 78, 96, 129, 136, 140, 152
- Church work, 41, 136, 148, 166, 216 *et seq.*

- Classification of stitches, 9, 175 *et seq.*
- Cloth, 125, 126, 159, 243
- Colour, 110, 208
- Colour gradation, 98, 114, 118
- Colour and outline, 146, 185
- Combination of stitches, 182
- Coptic embroidery, 12, 226
- Coptic tapestry, 2
- Coral, 166, 248
- Cord, 122
- Cord (couched), 128, 144, 178, 182
- [258] Cord (attachment of), 124
- Cotton, 243
- Couched cord, 128, 144, 178, 182
- Couched gold, 131 *et seq.*, 182
- Couched outline, 146
- Couching, 22, 114, 120, 121, 122 *et seq.*, 244
- Couching (reverse), 130
- Counterchange, 154
- Cretan embroidery, 12
- Cretan-stitch, 61 (*See also Ladder-stitch*)
- Crewel, 244
- Crewel-stitch, 26 *et seq.*, 83, 86, 103, 105, 178
- Crewel-stitch (surface), 86
- Crewel work, 26, 36, 37
- Cross-stitch, 12, 14, 16
- Crossed buttonhole-stitch, 56
- Cushion-stitch, 20, 21
- Cut-work, 156

- Damask, 243, 244
- Damping, 254, 255
- Darning, 8, 22, 83, 90, 106 *et seq.*, 178, 179
- Darning (Japanese), 86
- Darning (surface), 84
- Design, 150, 219, 233 *et seq.*

- Design, traditional, 238, 240
- Design and stitch, 10, 238
- Designer and embroiderer, 232, 233
- Diapers, 87, 88, 108, 132, 134, 210
- Direction of stitch, 92, 95, 108, 114, 136, 190, 208 *et seq.*
- Double darning, 106
- Double thread, 253
- Dovetail-stitch, 103, 104 (*See also Embroidery and Plumage Stitches*)
- Drawing with the needle, 192, 194, 196, 199, 211
- Drawn work, 2, 4

- Eastern embroidery. (*See Oriental*)
- Effect and stitch, 36, 78
- Eighteenth century embroidery, 220, 246
- Embroidery and painting, 201, 202
- Embroidery-stitch, 103 (*See also Plumage-stitch*)
- English embroidery, 34, 36, 169

- Feather-stitch, 62 *et seq.*, 83, 100, 178
- Felt, 243
- Fifteenth century embroidery, 24, 164
- Figure work, 116, 169, 190, 198 *et seq.*
- Filling-in patterns, 24
- Filo-floss, 164, 244
- Filoselle, 124, 144, 244
- Fishbone, 21, 47, 51
- Flax thread, 164, 244
- Flemish embroidery, 142, 200
- Flesh, 204, 206
- Florentine-stitch, 18, 21 (*See also Cushion stitch*)
- [259] Floss, 95, 114, 116, 118, 120, 244
- Form and stitch, 44, 47, 100, 118, 176, 211, 253
- Framing work, 251
- French embroidery, 88, 245
- French floss, 244

- French knots, 77, 129, 150, 178, 244

- Geometric pattern, 225
- German embroidery, 110, 125, 126, 156, 185, 226
- German knot-stitch, 72
- Gobelin-stitch, 18
- Gold, 210, 222, 245
- Gold (couched), 131 *et seq.*, 182
- Gold (raised), 134, 136, 165
- Gold thread, 131, 245
- Gold tinted by couching stitches, 142
- Gold wire, 169, 248

- Half-cross-stitch, 20
- Heraldic embroidery, 156
- Herringbone-stitch, 8, 22, 47 *et seq.*, 83, 178, 182
- Hildesheim cope (the), 126
- Hungarian embroidery, 2
- Hungarian stitch, 18

- Indian embroidery, 41, 46, 61, 95, 98, 154, 169, 222, 248
- Indian herring-bone, 48
- Inlay, 153
- Interlacing stitches, 83
- Italian embroidery, 22, 24, 37, 46, 138
- Italian embroidery (Renaissance), 22, 41, 120, 142, 154, 199

- Japanese darning, 86, 87
- Japanese embroidery, 80
- Japanese gold, 245
- Jewels, 165, 248

- Knot stitches, 72 *et seq.*, 182

- Lace, 1, 2

- Lace stitches, 84 *et seq.*
- Ladder-stitch, 59, 61, 182
- Laid-work, 112 *et seq.*, 162, 178
- Leather, 248
- Leather on velvet, 150
- Length of stitch, 96, 100
- Limitations of embroidery, 240
- Line work, 176, 178
- Linen, 164, 243
- Linen (embroidery on), 24
- Long-and-short-stitch, 36, 98, 100, 178, 190, 192

- Magic-stitch, 41
- Material (influence of on stitch), 12, 13, 16, 18, 24, 88, 91
- Materials, 242 *et seq.*
- Mechanical embroidery, 225
- Mediæval work, 92, 136, 140, 190
- Milanese-stitch, 18
- Modelling, 222
- [260] Modest work, 230, 231
- Moorish-stitch, 18, 21
- Morocco embroidery, 152

- Needle (tambour), 38, 245
- Needle pictures, 201
- Needles, 250
- Net passing, 86

- Old English Knot-stitch, 75
- Opus Anglicanum, 9
- Oriental embroidery, 2, 22, 61, 92, 112, 136, 140, 153, 226
- Oriental stitch, 66 *et seq.*, 83, 178, 182
- Originality, 234
- Outline, 22, 77, 108, 146, 158, 178, 184, 185 *et seq.*
- Outline (couched), 126, 128, 146
- Outline (double), 146, 185, 186

- Outline (stepped), 16, 24
- Outline (voided), 96, 187
- Outline embroidery, 138
- Outline stitch, 29, 30, 32, 86

- Padding, 159, 172
- Painting, 201, 202
- Parchment, 160, 168, 246
- Parisian-stitch, 18
- Patchwork, 156
- Pearls, 165, 166, 248
- Peasant work, 12, 13, 226
- Persian embroidery, 7, 24, 41, 174, 228
- Pictorial effect, 198, 199, 201
- Pictures (tent-stitch), 14, 20
- Pierce, 132
- Pins, 146, 250
- Plait-stitch, 21
- Plate, 245
- Plumage-stitch, 62, 100, 103, 178, 179, 192, 212
- Preciousness, 198
- Purl, 245
- Purse silk, 116, 162

- Quilting, 172 *et seq.*

- Raised gold, 134, 136, 165 *et seq.*
- Raised work, 134, 136, 159 *et seq.*
- Relief, 159 *et seq.*, 166, 168, 169, 172, 222
- Renaissance embroidery, 41, 92, 142, 154, 166
- Renewing ground, 126
- Reverse-couching, 130
- Ribbon, 150, 246
- Ribbon work, 246
- Roll-stitch, 75 (*See also Bullion-stitch*)
- Roman satin, 243

- Rope-stitch, 71 *et seq.*, 178
- Running, 83, 106, 179

- Satin, 243
- Satin "de luxe", 243
- Satin on velvet, 150
- Satin-stitch, 24, 91 *et seq.*, 103, 112, 128, 158, 160, 162, 175, 178, 182, 192, 206, 212, 245
- Satin-stitch (surface), 98, 182
- [261] Satin-stitch in the making, 91
- Scissors, 250
- Serge, 243
- Seventeenth century embroidery, 14, 166
- Shaded silk, 246
- Shading, 34, 176, 188 *et seq.*
- Silk, 146, 243
- Silk (tussah), 244
- Silk (twisted), 95, 124, 125
- Silk on silk, 150
- Silks, 244
- Silver, 135, 138, 166
- Simplicity, 180, 236, 238
- Simplicity (a plea for), 225 *et seq.*
- Sixteenth century embroidery, 22, 120, 125, 142, 185, 199
- Solid chain-stitch, 43, 44
- Solid crewel-stitch, 32, 34
- Soudanese embroidery, 112
- Spangles, 169, 248
- Spanish embroidery, 129, 142, 154, 166, 185
- Spanish-stitch, 18, 22 (*See also Plait-stitch*)
- Split-stitch, 38, 100, 105, 114, 179, 190, 196, 222
- Spot-stitch, 30
- Stem-stitch, 32
- Stems, 95
- Stepped outline, 16, 24
- Stiletto, 174

- Stitch (definition of), 11
- Stitch adaptation, 103, 188, 253
- Stitch and effect, 36, 78
- Stitch and form, 44, 47, 100, 118, 176, 211, 253
- Stitch and stuff, 12, 13, 16, 18, 24, 88, 91
- Stitch groups, 9, 175 *et seq.*
- Stitch names, 8, 9
- Stitch patterns, 87, 88
- Stitch and design, 10, 238
- Stitches, 7
- Stitching over stitching, 215
- Stretching work, 251, 254
- String, 159, 160, 162
- Stroke-stitch, 16
- Stuffs, 242
- Surface crewel-stitch, 86
- Surface darning, 84
- Surface satin-stitch, 98, 182
- Surface stitches, 84
- Syon cope (the), 7, 130, 210

- Tailors' buttonhole, 56
- Tambour, 245
- Tambour frame, 44
- Tambour needle, 38, 245
- Tambour stitch, 38
- Tambour work, 44, 194
- Tapestry, 1, 2, 4, 143, 220
- Tapestry-stitch, 53
- Tendrils, 130
- Tent-stitch, 14, 18
- Thimble, 250
- Thread, 244
- Traditional design, 238, 240
- Transferring design, 251
- Turkish embroidery, 22

- Tussah silk, 244
- Twisted silk, 95, 124, 125

- Underlay, 159, 160, 165
- Unpicking, 253

- [262] Vandyke chain, 42
- Variety of method, 148, 158
- Variety of stitch, 180 *et seq.*
- Velvet, 150, 222
- Venetian embroidery, 138
- Voiding, 96, 187

- Weaving, 2
- White on white, 162, 230
- Wool. (*See Crewel*)
- Woollen stuffs, 243

THE END.

BRADBURY, AGNEW, & CO. LD., PRINTERS, LONDON AND TONBRIDGE.

*A LIST OF STANDARD BOOKS
ON
ORNAMENT & DECORATION,
INCLUDING
FURNITURE, WOOD-CARVING, METAL WORK,
&c.,
PUBLISHED BY
B. T. BATSFORD,
94, HIGH HOLBORN, LONDON, W.C.*

WINDOWS.—A BOOK ABOUT STAINED AND PAINTED GLASS. By Lewis F. Day. Containing 410 pages, including 50 full-page Plates, and upwards of 200 Illustrations in the text, all of Old Examples. Large 8vo, cloth gilt. Price 21s. net.

"Contains a more complete popular account—technical and historical—of stained and painted glass than has previously appeared in this country."—*The Times.*

"The book is a masterpiece in its way ... amply illustrated and carefully printed; it will long remain an authority on its subject."—*The Art Journal.*

"All for whom the subject of stained glass possesses an interest and a charm, will peruse these pages with pleasure and profit."—*The Morning Post.*

"Mr. Day has done a worthy piece of work in more than his usual admirable manner ... the illustrations are all good and some the best black-and-white drawings of stained glass yet produced."—*The Studio.*

Now Published, the most handy, useful, and comprehensive work on the subject.

ALPHABETS, OLD AND NEW. Containing 150 complete Alphabets, 30 Series of Numerals, Numerous Facsimiles of Ancient Dates. Selected and arranged by Lewis F. Day. Preceded by a short account of the Development of the Alphabet. With Modern Examples specially Designed by *Walter Crane, Patten Wilson, A. Beresford Pite*, the Author, and others. Crown 8vo, art linen. Price 3s. 6d. net.

"Mr. Day's explanation of the growth of form in letters is particularly valuable.... Many excellent alphabets are given in illustration of his remarks." — *The Studio.*

"Everyone who employs practical lettering will be grateful for 'Alphabets, Old and New.' Mr. Day has written a scholarly and pithy introduction, and contributes some beautiful alphabets of his own design." — *The Art Journal.*

"A practical resumé of all that is to be known on the subject, concisely and clearly stated." — *St. James' Gazette.*

"It goes without saying that whatever Mr. Batsford publishes and Mr. Day has to do with is presented in a good artistic form, complete, and wherever that is possible, graceful." — *The Athenæum.*

ARCHITECTURE AMONG THE POETS. By H. Heathcote Statham. With 13 Illustrations. Square 8vo, artistically bound. Price 3s. 6d. net.

"This little work does for architecture in relation to English poetry what Mr. Phil Robinson has done for the birds and beasts. The poet's appreciation of architecture is a delightful subject with which Mr. Statham has become infected, not only illustrating his points with quotations and his judgments with his reasons, but the whole with a series of fanciful or suggestive sketches which add considerably to the attractiveness of the book." — *The Magazine of Art.*

THE DECORATION OF HOUSES. By Edith Wharton and Ogden Codman, Architect. 204 pages of text, with 56 full-page Photographic Plates of Views of Rooms, Doors, Ceilings, Fireplaces, various pieces of Furniture, &c., from the Renaissance period. Large square 8vo, cloth gilt, price 12s. 6d. net.

This volume, written by an American Lady Artist, and an Architect, describes and illustrates in a very interesting way the Decorative treatment of Rooms during the Renaissance period, and deduces principles for the decoration, furnishing, and arrangements of Modern Houses.

"... has illustrations which are beautiful ... because they illustrate the sound and simple principle of decoration which the authors put forward.... The book is one which should be in the library of every man and woman of means, for its advice is characterised by so much common sense as well as by the best of taste." — *The Queen.*

THE HISTORIC STYLES OF ORNAMENT. Containing 1,500 examples from all countries and all periods, exhibited on 100 Plates, mostly printed in gold and colours. With historical and descriptive text translated from the German of H. Dolmetsch. Folio, handsomely bound in cloth, gilt, price £1 5s. net.

This work has been designed to serve as a practical guide for the purpose of showing the development of Ornament, and the application of colour to it in various countries through the epochs of history. The work illustrates not only Flat Ornament, but also many Decorative Objects, such as Metal-Work, Pottery and Porcelain, Lace, Enamel, Mosaic, Illumination, Stained Glass, Jewellery, Bookbinding, &c., showing the application of Ornament to Industrial Art.

Just Published.

A MANUAL OF HISTORIC ORNAMENT, being an Account of the Development of Architecture and the Historic Arts, for the use of Students and Craftsmen. By Richard Glazier, A.R.I.B.A., Headmaster of the Manchester School of Art. Containing 42 Plates and 100 Illustrations in the text. Demy 8vo, cloth. Price 5s.

The object of this book is to furnish students with a concise account of Historic Ornament, in which the rise of each style is noted, and its characteristic features illustrated. It contains upwards of 400 subjects drawn by the author, and includes examples of Architectural Detail and Plastic Ornament, Pottery, Textile Fabrics, Glass, Metal-work, Mosaic, Painted Faïence, &c., &c. of various countries.

A MANUAL OF PRACTICAL INSTRUCTION IN THE ART OF BRASS REPOUSSÉ FOR AMATEURS. By Gawthorp (Art Metal Worker to H.R.H. the Prince of Wales). Second and enlarged Edition. With 32 Illustrations, many from photographs of executed designs. Crown 8vo, in wrapper. Price 1s. net.

OLD CLOCKS AND WATCHES AND THEIR MAKERS. By F. J. Britten, Secretary of the Horological Institute. Being an Account of the History of Clocks and Watches, their Mechanism and Ornamentation, to which is appended a List of 8,000 Old Makers, with descriptive Notes. Containing over 400 Illustrations, many reproduced from photographs, of choice and curious examples, of Clocks and Watches of the past in England and abroad, including the finely-ornamented Bracket Clocks of the XVIIth Century, with their ingenious mechanism, and the tall and elegant cases of the XVIIIth Century, also a selection of Portraits of the most renowned Masters of the Clockmaker's Art. 512 pages. Demy 8vo, cloth, gilt. Price 10s. net.

KING RENÉ'S HONEYMOON CABINET. A Monograph. By *John P. Seddon*, Architect. Illustrated by 10 photographic reproductions of the Cabinet, and the Panels, painted by the late Sir E. Burne Jones, *Dante Gabriel Rossetti*, and *Ford Madox Brown*. With a chapter on the Hereditary Earls of Anjou, by G. H. Birch, F.S.A. Large 8vo, cloth, price 5s. net.

This interesting little work has been issued by the author to make known and commemorate some early designs by the celebrated artists. Very few copies are printed for sale.

A small remainder, just reduced in price.

ANIMALS IN ORNAMENT. By Professor G. Sturm. Containing 30 large collotype plates, printed in tint, of designs suitable for Friezes, Panels, Borders, Wall-papers, Carving, and all kinds of Surface Decoration, &c. Large folio in portfolio, price 18s. net (published £1 10s.).

A new and useful series of clever designs, showing how animal forms may be adapted to decorative purposes with good effect.

A HISTORY OF DESIGN IN PAINTED GLASS. — From the Earliest Times to the end of the Seventeenth Century. By N. H. J. Westlake, F.S.A. Containing 467 illustrations with historical text. Four volumes, small folio, cloth, price £5 10s., net £4 8s.

Very few copies remain for sale of this valuable work.

MR. LEWIS F. DAY'S TEXT BOOKS OF ORNAMENTAL DESIGN.

SOME PRINCIPLES OF EVERY-DAY ART.—Introductory Chapters on the Arts not Fine. Forming a Prefatory Volume to the Series of Text Books. Second Edition, revised, containing 70 Illustrations (Third Thousand). Crown 8vo, art linen, price 3s. 6d., net 3s.

"Authoritative as coming from a writer whose mastery of the subjects is not to be disputed, and who is generous in imparting the knowledge he acquired with difficulty. Mr. Day has taken much trouble with the new edition."—*Architect.*

"A good artist, and a sound thinker, Mr. Day has produced a book of sterling value."—*Magazine of Art.*

THE ANATOMY OF PATTERN.—Containing: I. Introductory. II. Pattern Dissections. III. Practical Pattern Planning. IV. The "Drop" Pattern. V. Skeleton Plans. VI. Appropriate Pattern. Fourth Edition (Ninth Thousand), revised, with 41 full-page Illustrations. Crown 8vo, art linen, price 3s. 6d., net 3s.

"... There are few men who know the science of their profession better or can teach it as well as Mr. Lewis Day; few also who are more gifted as practical decorators; and in anatomising pattern in the way he has done in this manual—a way beautiful as well as useful—he has performed a service not only to the students of his profession, but also to the public."—*Academy.*

THE PLANNING OF ORNAMENT.—Containing: I. Introductory. II. The Use of the Border. III. Within the Border. IV. Some Alternatives in Design. V. On the Filling of the Circle and other Shapes. VI. Order and Accident. Third Edition (Fifth Thousand), further revised, with 41 full-page Illustrations, many of which have been redrawn. Crown 8vo, art linen, price 3s. 6d., net 3s.

"Contains many apt and well-drawn illustrations; it is a highly comprehensive, compact, and intelligent treatise on a subject which is more difficult to treat than outsiders are likely to think. It is a capital little book, from which no tyro (it is addressed to improvable minds) can avoid gaining a good deal."—*Athenæum.*

THE APPLICATION OF ORNAMENT.—Containing: I. The Rationale of the Conventional. II. What is Implied by Repetition. III. Where to Stop in Ornament. IV. Style and Handicraft. V. The Teaching of the Tool. VI. Some Superstitions. Third Edition (Sixth Thousand), further revised, with 48 full-page Illustrations and 7 Woodcuts in the text. Crown 8vo, art linen, price 3s. 6d., net 3s.

"A most worthy supplement to the former work, and a distinct gain to the art student who has already applied his art knowledge in a practical manner, or who hopes yet to do so."—*Science and Art.*

ORNAMENTAL DESIGN.—Comprising the above Three Books, "Anatomy of Pattern," "Planning of Ornament," and "Application of Ornament," handsomely bound in one volume, cloth gilt, price 10s. 6d., net 8s. 6d.

NATURE IN ORNAMENT.—With 123 full-page Plates and 192 Illustrations in the text. Third Edition (Fifth Thousand). Thick crown 8vo, in handsome cloth binding, richly gilt, price 12s. 6d., net 10s.

Contents: I. Introductory. II. Ornament in Nature. III. Nature in Ornament. IV. The Simplification of Natural Forms. V. The Elaboration of Natural Forms. VI. Consistency in the Modification of Nature. VII. Parallel Renderings. VIII. More Parallels. IX. Tradition in Design. X. Treatment. XI. Animals in Ornament. XII. The Element of the Grotesque. XIII. Still Life in Ornament. XIV. Symbolic Ornament.

"Amongst the best of our few good ornamental designers is Mr. Lewis F. Day, who is the author of several books on ornamental art. 'Nature in Ornament' is the latest of these, and is probably the best. The treatise should be in the hands of every student of ornamental design. It is profusely and admirably illustrated, and well printed."—*Magazine of Art.*

"A book more beautiful for its illustrations, or one more helpful to Students of Art, can hardly be imagined."—*Queen.*

A HANDBOOK OF ORNAMENT.—With 300 Plates, containing about 3,000 Illustrations of the Elements and Application of Decoration to Objects. By F. S. Meyer, Professor at the School of Applied Art, Karlsruhe. Third English Edition, revised by Hugh Stannus, Lecturer on Applied Art at the Royal College of Art, South Kensington. Thick 8vo, cloth gilt, gilt top, price 12s. 6d., net 10s.

"A Library, a Museum, an Encyclopædia and an Art School in one. To rival it as a book of reference, one must fill a bookcase. The quality of the drawings is unusually high, the choice of examples is singularly good.... The work is practically an epitome of a hundred Works on Design." — *Studio.*

"The author's acquaintance with ornament amazes, and his three thousand subjects are gleaned from the finest which the world affords. As a treasury of ornament drawn to scale in all styles, and derived from genuine concrete objects, we have nothing in England which will not appear as poverty-stricken as compared with Professor Meyer's book." — *Architect.*

"The book is a mine of wealth even to an ordinary reader, while to the Student of Art and Archæology it is simply indispensable as a reference book. We know of no one work of its kind that approaches it for comprehensiveness and historical accuracy." — *Science and Art.*

A HANDBOOK OF ART SMITHING. — For the use of Practical Smiths, Designers and others, and in Art and Technical Schools. By F. S. Meyer, Author of "A Handbook of Ornament." Translated from the Second German Edition. With an Introduction by J. Starkie Gardner. Containing 214 Illustrations. Demy 8vo, cloth, price 6s., net 5s.

Both the Artistic and Practical Branches of the subject are dealt with, and the Illustrations give selected Examples of Ancient and Modern Ironwork. The Volume thus fills the long-existing want of a Manual on Ornamental Ironwork, and it is hoped will prove of value to all interested in the subject.

"Charmingly produced.... It is really a most excellent manual, crowded with examples of ancient work, for the most part extremely well selected." — *The Studio.*

"Professor Meyer's work is a useful historical manual on art smithing, based on a scientific classification of the subject, that will be of service to all smiths, designers, and students of technical and art schools. The illustrations are well drawn and numerous." — *Building News.*

Published with the Sanction of the Science and Art Department.

FRENCH WOOD CARVINGS FROM THE NATIONAL MUSEUMS.—A Series of Examples printed in Collotype from Photographs specially taken from the Carvings direct. Edited by Eleanor Rowe. Part I.: Late 15th and Early 16th Century Examples; Part II.: 16th Century Work; Part III.: 17th and 18th Centuries. The Three Series Complete, each containing 18 large folio Plates, with descriptive letterpress. Folio, in portfolios, price 12s. each net; or handsomely bound in one volume, £2 5s. net.

"Students of the Art of Wood Carving will find a mine of inexhaustible treasures in this series of illustrations of French Wood Carvings.... Each plate is a work of art in itself; the distribution of light and shade is admirably managed, and the differences in relief are faithfully indicated, while every detail is reproduced with a clearness that will prove invaluable to the student. Sections are given with several of the plates."—*The Queen.*

"Needs only to be seen to be purchased by all interested in the craft, whether archæologically or practically."—*The Studio.*

HINTS ON WOOD CARVING FOR BEGINNERS.—By Eleanor Rowe. Fourth Edition, revised and enlarged, Illustrated. 8vo, sewed, price 1s. in paper covers, or bound in cloth, price 1s. 6d.

"The most useful and practical small book on wood-carving we know of."—*Builder.*

"... Is a useful little book, full of sound directions and good suggestions."—*Magazine of Art.*

HINTS ON CHIP CARVING.—(Class Teaching and other Northern Styles.) By Eleanor Rowe. 40 Illustrations. 8vo, sewed, price 1s. in paper covers, or in cloth, price 1s. 6d.

"A capital manual of instruction in a craft that ought to be most popular."—*Saturday Review.*

DETAILS OF GOTHIC WOOD CARVING.—Being a Series of Drawings from original work of the Fourteenth and Fifteenth Centuries. By Franklyn A. Crallan. Containing 34 large Photolithographic Plates, with introductory and descriptive text. Large 4to, in handsome cloth portfolio, or bound in cloth gilt, price 28s., net 22s.

"The examples are carefully drawn to a large size ... well selected and very well executed." — *The Builder.*

PROGRESSIVE STUDIES AND DESIGNS FOR WOOD-CARVERS. By Miss E. R. Plowden. With a Preface by Miss Rowe. Consisting of five large folding sheets of Illustrations (drawn full size), of a variety of objects suitable for Wood Carving. With descriptive text. Second Edition, enlarged. 4to, in portfolio. Price 5s. net.

ANCIENT WOOD AND IRONWORK IN CAMBRIDGE. — By W. B. Redfarn, the Letterpress by John Willis Clark. 29 folio Lithographed Plates drawn to a good scale. Cloth gilt, a handsome volume, price 10s. 6d., net 8s. 6d.

This Work, giving an interesting and useful series of Examples, is but little known. Very few copies remain.

HEPPLEWHITE'S CABINET-MAKER AND UPHOLSTERER'S GUIDE; or Repository of Designs for every article of Household Furniture in the newest and most approved taste. A complete facsimile reproduction of this rare work, containing nearly 300 charming Designs on 128 Plates. Small folio, bound in speckled cloth, gilt, old style, price £2 10s. net. (1794.) *Original copies when met with fetch from £17 to £18.*

"A beautiful replica, which every admirer of the author and period should possess." — *Building News.*

CHIPPENDALE'S THE GENTLEMAN AND CABINET-MAKER'S DIRECTOR. — A complete facsimile of the 3rd and rarest Edition, containing 200 Plates of Designs of Chairs, Sofas, Beds and Couches, Tables, Library Book Cases, Clock Cases, Stove Grates, &c., &c. Folio, strongly bound in half-cloth, price £3 15s. net. (1762.)

SHERATON'S CABINET-MAKER And UPHOLSTERER'S DRAWING-BOOK. — A complete Facsimile Reproduction of the scarce Third Edition. With the rare Appendix and Accompaniment complete. Containing in all 434 pages and 122 Plates. 4to, cloth, price £2 10s. net.

EXAMPLES OF OLD FURNITURE, English and Foreign. Drawn and described by Alfred Ernest Chancellor. Containing 40 Photolithographic Plates exhibiting some 100 examples of Elizabethan, Stuart, Queen Anne, Georgian and Chippendale furniture; and an interesting variety of Continental work. With historical and descriptive notes. Large 4to, gilt, price £1 5s., net £1 1s.

"In publishing his admirable collection of drawings of old furniture, Mr. Chancellor secures the gratitude of all admirers of the consummate craftsmanship of the past. His examples are selected from a variety of sources with fine discrimination, all having an expression and individuality of their own—qualities that are so conspicuously lacking in the furniture of our own day. It forms a very acceptable work." — *The Morning Post*.

FURNITURE AND DECORATION IN ENGLAND DURING THE XVIIIth CENTURY. — By J. Aldam Heaton. Two volumes, each of two parts, bound in four, large folio, cloth, price £7 net. Containing upwards of 150 plates of photographic reproductions from the published designs of R. & J. Adam, Chippendale, Hepplewhite, Sheraton, Shearer, Pergolesi, Cipriani, Darly, Johnson, Richardson, and all great English designers and cabinet-makers of the period.

This work forms an encyclopædic and almost inexhaustible treasury of reference for all Furniture Designers, Painters, Interior Decorators, Cabinet-makers, &c., since no artist of importance is unrepresented, and a fair selection is in every case given of his work.

REMAINS OF ECCLESIASTICAL WOOD-WORK. — A Series of Examples of Stalls, Screens, Book-Boards, Roofs, Pulpits, &c., containing 21 Plates beautifully engraved on Copper, from drawings by T. Talbot Bury, Archt. 4to, half-bound, price 10s. 6d., net 8s. 6d.

FLAT ORNAMENT: A Pattern Book for Designers of Textiles, Embroideries, Wall Papers, Inlays, &c., &c. — 150 Plates, some printed in Colours, exhibiting upwards of 500 Historical Examples of Textiles, Embroideries, Paper Hangings, Tile Pavements, Intarsia Work, &c. With some Designs by Dr. Fischbach. Imperial 4to boards, cloth back, price £1 5s., net 20s.

EXAMPLES OF ENGLISH MEDIÆVAL FOLIAGE AND COLOURED DECORATION. — By Jas. K. Colling, Architect, F.R.I.B.A. Taken from Buildings of the XIIth to the XVth Century. Containing

76 Lithographic Plates, and 79 Woodcut Illustrations, with Text. Royal 4to, cloth, gilt top, price 18s., net 15s. (published at £2 2s.)

PLASTERING—PLAIN AND DECORATIVE. A Practical Treatise on the Art and Craft of Plastering and Modelling. Including full descriptions of the various Tools, Materials, Processes and Appliances employed. With over 50 full-page Plates, and about 500 smaller Illustrations in the Text. By William Millar. With an Introduction, treating of the History of the Art, by G. T. Robinson, F.S.A. Thick 4to, cloth, containing 600 pages of text, price 18s. net.

"This new and in many senses remarkable treatise ... unquestionably contains an immense amount of valuable first-hand information.... 'Millar on Plastering' may be expected to be the standard authority on the subject for many years to come.... A truly monumental work."—*The Builder.*

A GRAMMAR OF JAPANESE ORNAMENT AND DESIGN.—Illustrated by 65 Plates, many in Gold and Colours, representing all Classes of Natural and Conventional Forms, drawn from the Originals, with introductory, descriptive, and analytical text. By T. W. Cutler, F.R.I.B.A. Imperial 4to, in elegant cloth binding, price £2 6s., £1 18s. net.

DECORATIVE WROUGHT IRONWORK OF THE 17th AND 18th CENTURIES.—By D. J. Ebbetts. Containing 16 large Lithographic Plates, illustrating 70 English examples of Screens, Grilles, Panels, Balustrades, &c. Folio, boards, cloth back, price 12s. 6d., net 10s.

A Facsimile reproduction of one of the rarest and most remarkable Books of Designs ever published in England.

A NEW BOOKE OF DRAWINGS OF IRONWORKE.—Invented and Desined by John Tijou. Containing severall sortes of Iron Worke, as Gates, Frontispieces, Balconies, Staircases, Pannells, &c., of which the most part hath been wrought at the Royall Building of Hampton Court, &c. All for the use of them that worke iron in perfection and with art. (Sold by the author in London, 1693.) Containing 20 folio Plates. With Introductory Note and Descriptions of the Plates by J. Starkie Gardner. Folio, bound in boards, old style, price 25s. net.

Only 150 copies were printed for England, and very few now remain. An original copy is priced at £48 by Mr. Quaritch, the renowned bookseller.

JAPANESE ENCYCLOPÆDIA OF DESIGN.

Book I.—Containing over 1,500 engraved curios, and most ingenious Geometric Patterns of Circles, Medallions, &c., comprising Conventional Details of Plants, Flowers, Leaves, Petals, also Birds, Fans, Animals, Key Patterns, &c., &c. Oblong 12mo, fancy covers, price 2s. net.

Book II.—Containing over 600 most original and effective Designs for Diaper Ornament, giving the base lines to the design, also artistic Miniature Picturesque Sketches. Oblong 12mo, price 2s. net.

These books exhibit the varied charm and originality of conception of Japanese Ornament, and form an inexhaustible field of design.

A DELIGHTFUL SERIES OF STUDIES OF BIRDS, in most Characteristic and Life-like Attitudes, surrounded with appropriate Foliage and Flowers.—By the celebrated Japanese Artist, Bairei Kono. In three Books, 8vo, each containing 36 pages of highly artistic and decorative Illustrations, printed in tints. Bound in fancy paper covers, price 10s. net.

"In attitude and gesture and expression, these Birds, whether perching or soaring, swooping or brooding, are admirable."—*Magazine of Art.*

A NEW SERIES OF BIRD AND FLOWER STUDIES. BY Watanabe Sietei, the acknowledged leading living Artist in Japan. In 3 Books, containing numerous exceedingly Artistic Sketches in various tints, 8vo, fancy covers. Price 10s. net.

ARTISTS' SKETCH BOOKS.—A SERIES OF FIVE VOLUMES.—Vol. I.: Birds, Flowers, and Plants, drawn in a Decorative Spirit. Vol. II.: Sketches of Insects, Plants, &c., drawn for Designers. Vol. III.: Drawings of Fishes and Marine Animals. Vol. IV.: Natural Scenery, Landscapes, &c. Vol. V.: Scenes from Japanese Life, &c. 8vo, fancy covers. 7s. 6d. net.

THE ARCHITECTURE OF THE RENAISSANCE IN ITALY.—A General View for the Use of Students and Others. By W. J. Anderson, A.R.I.B.A., Director of Architecture, Glasgow School of Art. Second Edition, revised and enlarged. Containing 64 full-page Plates, mostly reproduced from Photographs, and 100 Illustrations in text. Large 8vo, cloth gilt, price 12s. 6d. net.

"A delightful and scholarly work ... very fully illustrated."—*Journal R.I.B.A.*

"It is the work of a scholar taking a large view of his subject.... The book affords easy and intelligible reading, and the arrangement of the subject is excellent, though this was a matter of no small difficulty."—*The Times.*

"Should rank amongst the best architectural writings of the day."—*The Edinburgh Review.*

"We know of no book which furnishes such information and such illustrations in so compact and attractive a form. For greater excellence with the object in hand there is not one more perspicuous."—*The Building News.*

A HISTORY OF ARCHITECTURE for the Student, Craftsman and Amateur.—Being a Comparative View of the Historical Styles from the Earliest Period. By Banister Fletcher, F.R.I.B.A., Professor of Architecture in King's College, London, and B. F. Fletcher, A.R.I.B.A. Containing 300 pages, with 115 Collotype Plates, mostly from large Photographs, and other Illustrations in the text. Third Edition, revised. Cr. 8vo, cloth gilt, price 12s. 6d., net 10s.

"We shall be amazed if it is not immediately recognised and adopted as *par excellence* the student's manual of the history of architecture."—*The Architect.*

"The general reader will read the book with not less profit than the student, and will find in it quite as much as he is likely to retain in his memory, and the architectural student in search of any particular fact will readily find it in this most methodical work.... As complete as it well can be."—*The Times.*

"As a synopsis of architectural dates and styles, Professor Banister Fletcher's work will fill a void in our literature, and become a most useful manual." — *The Building News.*

THE ORDERS OF ARCHITECTURE: GREEK, ROMAN AND ITALIAN. — Edited with Notes by R. Phené Spiers, F.S.A., F.R.I.B.A. Third Edition, revised and enlarged, containing 26 Plates. 4to, cloth, price 10s. 6d., net 8s. 6d.

"A most useful work for architectural students.... Mr. Spiers has done excellent service in editing this work, and his notes on the plates are very appropriate and useful." — *British Architect.*

RENAISSANCE ARCHITECTURE AND ORNAMENT IN SPAIN. — A Series of Examples selected from the purest executed between the years 1500-1560. By Andrew N. Prentice, A.R.I.B.A. Containing 60 beautiful Plates, reproduced by Photo-lithography and Photo Process from the author's drawings, of Perspective Views and Geometrical Drawings, and details, in Stone, Wood, and Metal. With short descriptive text. Folio, handsomely bound in cloth gilt, price £2 10s., net £2 2s.

"For the drawing and production of this book one can have no words but praise.... It is a pleasure to have so good a record of such admirable Architectural Drawing, free, firm and delicate." — *British Architect.*

B. T. BATSFORD, 94, HIGH HOLBORN, LONDON.

www.ingramcontent.com/pod-product-compliance
Lightning Source LLC
Chambersburg PA
CBHW031615210526
45464CB00004B/1585